# Nature's Hidden Force

## Joining Spirituality with Science

# Nature's Hidden Force

## Joining Spirituality with Science

George Land and
Beth Jarman

HUMANIST PRESS

WASHINGTON, DC

Humanist Press
1777 T Street NW
Washington, DC, 20009
(202) 238-9088

www.humanistpress.com



Printed book ISBN: 978-0-931779-49-7
Ebook ISBN: 978-0-931779-50-3

Editors: Kim Kressaty and Luis Granados
Cover design: Lisa Zangerl
Cover Photo © Joerg Habermeier | Dreamstime.com

# Nature's Hidden Force

## Joining Spirituality with Science

George Land and
Beth Jarman

HUMANIST PRESS

WASHINGTON, DC

Humanist Press
1777 T Street NW
Washington, DC, 20009
(202) 238-9088

www.humanistpress.com



Printed book ISBN: 978-0-931779-49-7
Ebook ISBN: 978-0-931779-50-3

Editors: Kim Kressaty and Luis Granados
Cover design: Lisa Zangerl
Cover Photo © Joerg Habermeier | Dreamstime.com

*"Tis to detect the inmost force*
*Which binds the world and guides its course."*
(Faust, I lines 382-3)

**DEMOCRITUS (460-370 B. C)**

*"Everything existing in the Universe is the fruit of chance and necessity."*

# CONTENTS

**From the Publisher** . . . . . . . . . . . . . . . . . . . . . . . . . . . . . . . . viii

**Acknowledgements**. . . . . . . . . . . . . . . . . . . . . . . . . . . . . . . . ix

**Chapter One:** Searching for God. . . . . . . . . . . . . . . . . . . . . . . 3

**Chapter Two:** Northern Ireland. . . . . . . . . . . . . . . . . . . . . . . 17

**Chapter Three:** Getting to the Bottom of Things. . . . . . . . . 24

**Chapter Four:** The Invisible Box. . . . . . . . . . . . . . . . . . . . . 45

**Chapter Five:** New Possibilities. . . . . . . . . . . . . . . . . . . . . . 61

**Chapter Six:** Discovery and Revolution. . . . . . . . . . . . . . . . 85

**Chapter Seven:** Future Pull. . . . . . . . . . . . . . . . . . . . . . . . 103

**Chapter Eight:** Today and Yesterday's
Breakpoint Worlds. . . . . . . . . . . . . . . . . . . . . . . . . 118

**Chapter Nine:** Why, Why, Why?. . . . . . . . . . . . . . . . . . . . . 134

**Chapter Ten:** Uniting Spirit and Science. . . . . . . . . . . . . . . 145

**Appendix One:** Personal Acknowledgements. . . . . . . . . . . 158

**Appendix Two:** Current Entropy Definitions. . . . . . . . . . . 160

**Appendix Three:** Creativity and Probability. . . . . . . . . . . 163

**Appendix Four:** About the Authors. . . . . . . . . . . . . . . . . . 168

**Appendix Five:** Client Acknowledgments. . . . . . . . . . . . . . 170

**Your Turn**. . . . . . . . . . . . . . . . . . . . . . . . . . . . . . . . . . . . . . 173

**Index** . . . . . . . . . . . . . . . . . . . . . . . . . . . . . . . . . . . . . . . . 174

# FROM THE PUBLISHER

> Knowledge of the world is derived by observation, experimentation, and rational analysis. Humanists find that science is the best method for determining this knowledge as well as for solving problems and developing beneficial technologies. We also recognize the value of new departures in thought, the arts, and inner experience—each subject to analysis by critical intelligence.
>
> — Humanist Manifesto III

At Humanist Press, we strive to present a diverse range of viewpoints within the humanist tradition, to expand our readers' horizons.

What this book is *not* is "the official humanist version of physics," or anything akin to that.

Instead, this book should be thought of as a "new departure in thought ... subject to analysis by critical intelligence." It challenges some conventional ideas of physics, and suggests an interesting fusion of science and the set of feelings that many refer to as "spirituality."

If the ideas presented here make you re-think what you thought you already knew, if they spark debate or even outright disagreement, then we are fulfilling our mission. We encourage readers to exercise their critical intelligence and practice what the authors would call "Creative Connecting" by posting their comments (see page 173) for all to see.

# ACKNOWLEDGEMENTS

Many years, decades, ago we joined the search party. What a mixed bag the party was. Millennia ago, sandaled Persian philosophers trod the sands of the Silk Road; Athenian scholars argued in the Parthenon; alchemists in Germany played with colored fires; and others, sailors from Babylon or Carthaginian scribes, made their way to Africa to listen to ancient drums of wisdom. They searched the skies in Rome, looking and listening. For what? What could the sky tell them? What were they all wondering?

They all searched for some elusive knowledge.

*What's it all about?*

*What can the universe tell us?*

*Why are we here?*

Like those other searchers, we were waiting for the universe to speak to us – through the clues all those other searchers had found.

The concepts in this book came to us from them.

They came from all those other searchers, from the Tao, from Confucius, from Rumi, from Kant, from Descartes, from Mach and Spencer, from all those who have been whispering in our ears for so long. We have tried hard to listen and put the best words we could to their ideas. Whatever we have found, it is because over the millennia the great search party has been out probing and penetrating every corner they could find. All we could do is put their findings together in a new way, a way that could bind together the science of today with the ancient mythos of creation and love.

Two people in particular have been living with us over the years, looking over our shoulders and, we feel, urging us on: Albert Einstein and Ilya Prigogine. Einstein is, of course, well known; Ilya Prigogine, a more recent Nobel Laureate is, to us, at least his equal. Prigogine's ability to see the whole of nature and how that vast myriad of parts fits together is phenomenal — beyond anyone we have ever encountered. Their books, correspondence, notes, lectures and thinking have kept us going many times when we thought we had stepped over the edge. They have deeply touched our hearts and minds. We hope to meet them some day – though not too soon.

Please see Appendix One for the names of the people who have helped make this work possible.

# ACKNOWLEDGEMENTS

Many years, decades, ago we joined the search party. What a mixed bag the party was. Millennia ago, sandaled Persian philosophers trod the sands of the Silk Road; Athenian scholars argued in the Parthenon; alchemists in Germany played with colored fires; and others, sailors from Babylon or Carthaginian scribes, made their way to Africa to listen to ancient drums of wisdom. They searched the skies in Rome, looking and listening. For what? What could the sky tell them? What were they all wondering?

They all searched for some elusive knowledge.

*What's it all about?*

*What can the universe tell us?*

*Why are we here?*

Like those other searchers, we were waiting for the universe to speak to us – through the clues all those other searchers had found.

The concepts in this book came to us from them.

They came from all those other searchers, from the Tao, from Confucius, from Rumi, from Kant, from Descartes, from Mach and Spencer, from all those who have been whispering in our ears for so long. We have tried hard to listen and put the best words we could to their ideas. Whatever we have found, it is because over the millennia the great search party has been out probing and penetrating every corner they could find. All we could do is put their findings together in a new way, a way that could bind together the science of today with the ancient mythos of creation and love.

Two people in particular have been living with us over the years, looking over our shoulders and, we feel, urging us on: Albert Einstein and Ilya Prigogine. Einstein is, of course, well known; Ilya Prigogine, a more recent Nobel Laureate is, to us, at least his equal. Prigogine's ability to see the whole of nature and how that vast myriad of parts fits together is phenomenal — beyond anyone we have ever encountered. Their books, correspondence, notes, lectures and thinking have kept us going many times when we thought we had stepped over the edge. They have deeply touched our hearts and minds. We hope to meet them some day – though not too soon.

Please see Appendix One for the names of the people who have helped make this work possible.

# *Nature's Hidden Force*

## *Joining Spirituality with Science*

***All the world's major religions, with their emphasis on love, compassion, tolerance, and forgiveness can and do promote inner values. But the reality of the world today is that grounding ethics in religion is no longer adequate. That is why I am convinced that the time has come to find a way of thinking about spirituality and ethics beyond religion altogether.***

Dalai Lama in his book *Beyond Religion: Ethics for a Whole World*

CHAPTER ONE

# SEARCHING FOR GOD

---

One hundred and forty-eight years ago the scientific search for God came to an abrupt halt.

This development, following the publishing of Isaac Newton's laws of motion and gravitation in 1687, gave enormous impetus to the new idea of "science" as a way of gaining understanding about our universe and its laws. Science dramatically shifted the search for knowledge that could be based on verifiable facts.

A flood of scientific discoveries followed: planetary motion, cells, oxygen, fossils, chemistry, electricity, germs, enzymes, and many others. With the broad concept of gaining knowledge through science, suddenly real promise for answering the *big* questions was at hand. Belief in science grew.

There might even be a way of finding out more about the deity. The search for the nature of the deity after thousands of years of religious and philosophic speculation took on a new direction. Where did we come from? How was the universe designed? How does nature work? Could we discover nature's secrets and learn to apply them? What is the future of humanity? Some even believed that humans could answer all of these questions through science and there would be no need for revealed truth or what some called "blind faith." Many people expected the acquisition of scientific knowledge would confirm the details of their religious view, others thought that religion and science could exist side by side as one dealt with the spirit and the other with physical nature and they are altogether different domains.

In 1865, a new scientific law was discovered: a law as universal as gravity!

It is called the "Law of Entropy." It is part of a law called "The Second Law of Thermodynamics." It served as the crucial stop sign on the road on the way to finding God.

Jeremy Rifkin, in his book *Entropy* said, "All the foremost scientists agree that the law of entropy will remain the principal paradigm for the foreseeable future. Albert Einstein, the greatest scientist of our age, described it as the 'premier law of all of science.' The famous astrophysicist known for his work on relativity, Sir Arthur Eddington also referred to entropy as the 'supreme law of the entire universe.'"[1]

This law of entropy was thought to be so fundamental and important that the legendary scientist Sir Arthur Stanley Eddington, in his 1927 book *The Nature of the Physical World,* warned the scientific community about tinkering with it: "But if your theory is found to be against the second law of thermodynamics I can give you no hope; there is nothing for it but to collapse in deepest humiliation."

Even Albert Einstein supported the idea that entropy is not to be meddled with: "It is the only physical theory of universal content, which I am convinced, that within the

1. Jeremy Rifkin, *Entropy: A New World View* (New York: Viking Press, 1980) p. 6.

framework of applicability of its basic concepts will never be overthrown."[2]

It is certainly no wonder that scientists have steered clear of devoting their time and resources to questioning what was such a fundamental and paramount law. Arguing with leading authorities is no way to get ahead in science. There are easier challenges.

We want you to know why we have dedicated the last few decades of our lives to the study of this entropy section of the second law. In deepest humility, we are prepared to face humiliation as our findings do overthrow the basic concepts of entropy and actually form a powerful and now scientifically verifiable bridge between science and spirituality.

What does this fundamental "supreme" universal law say? Why would we want to overthrow this law? And what does that have to do with finding God?

**The Entropy Section of the Second Law of Thermodynamics:**

> *All processes in the universe manifest a tendency toward decay and disintegration, with a net increase in what is called the entropy, or state of randomness or disorder, of the system.*

This is called the entropy part of the second law of thermodynamics. Ultimately the universe will descend into cold chaos—what is called a "heat death." The universe is defined as a great machine running down and wearing out and in no case going in the other direction.

For 150 years this has been accepted as the basic design of the universe. If you are looking for God as a scientist you just

2. Albert Einstein, quoted in M.J. Klein, "Thermodynamics in Einstein's Universe," *Science*, 157 (1967) p. 509.

need to look for who made this basic law because that's what was laid down as the basic rule for how everything works.

In other words, the whole lot of the universe is designed to break down and fall apart. The law does not state how fast this will happen. But it will happen, as Murphy's Law puts it: "If anything can go wrong, it will."

As science accepts the current law, internationally honored scientist Richard Dawkins put it, "The universe we observe has precisely the properties we should expect if there is, at bottom, no design, no purpose, no evil, no good, nothing but blind, pitiless indifference."[3]

The famous physicist Steven Weinberg commented, "The more the universe seems comprehensible the more it also seems pointless. I regard the existence of this extraordinary universe as having a wonderful, awesome grandeur. It hangs there in all its glory, wholly and completely useless. To project onto it our human-inspired notion of purpose would, to my mind, sully and diminish it."[4]

If you point out to a scientist that if you look around you will see things getting better he or she will point out that the universe keeps books on this, and if it is getting better someplace, it is making up for it by getting worse somewhere else, and the force of entropy will surely make up for it by getting worse here later.

Edgar Morin, widely recognized as one of the most important European thinkers to emerge in the twentieth century, had this to say about entropy: "Isn't the growing complexity only a detour in the generalized disaster of a universe that is intrinsically and definitively tragic?"[5]

---

3. Richard Dawkins, *River Out of Eden* (New York: Basic Books, 1996), p.133.

4. Stephen Weinberg, *Dreams of a Final Theory: The Search for the Fundamental Laws of Nature* (New York: Vintage Books, 1993).

5. Edgar Morin, "La relation anthropo-bio-cosmique," in *L'Univers philosophique, tome 1 de Encyclopédie Philosophique Universelle* (Paris, PUF, 1989) p.89-90.

So, simply put, as science dug down as deeply as possible to find what Sir Arthur Eddington called the "supreme, the premier law,"[6] it found entropy, the most fundamental force that was continuously and universally creating disarray, turmoil and disorder.

The law exists. Look it up anywhere. Google it and you will get more than fifteen million references.

If we go by scientific findings, many people have been impelled to ask, "What kind of a god would create such a malevolent and pessimistic force driving the universe?" With such a supreme and basic law of nature, it is no wonder that science gave up the search for God. Ludwig Boltzmann, the physicist and mathematician who codified this depressing law in 1877, committed suicide in 1906.

We simply could not accept this law, notwithstanding Einstein and Eddington and the great body of scientists who accepted it. We could not believe that this "law" could be true, or that a deity would design such a force. We admit that we were thinking with our hearts, not our heads, but we could not deny the power of the message we were getting. God would not be malicious. There must be a better answer; possibly an even deeper force! If we could not find it, so be it, but surely the search was worth it. Against all advice, odds and resources available, we plowed ahead.

We found that people – and even scientists – do not like to talk about entropy. Understandably. When is the last time you heard it brought up in conversation or even on a science program on television? It required a serious commitment and quite a bit of digging for us to get at the answer, from following the latest discoveries in scientific research to searching notebooks hundreds of years old.

---

6. Sir Arthur Stanley Eddington, *The Nature of the Physical World* (New York: Macmillan, 1948) p. 74.

A few decades later … our hearts were vindicated.

As the jigsaw puzzle of facts came together we ultimately found that the basic law of entropy was and is *wrong*.

Not only is the law wrong, but the law of entropy actually hides the creative force driving the universe—and our lives!

So that's why we wrote this book!

The way it is wrong about nature's most fundamental law and force is truly exciting—it has the power to change our lives in extremely positive ways!

As we said, the original law stated that this entropy force is continually creating more and more disorder. With the discoveries made in science since this law was found, and examined over several decades of our research, we parted the curtains hiding the actual force behind everything and found that the *opposite* is actually happening.

Not *disordering* is happening, but an extraordinarily special kind of *ordering* is occurring instead.

A great mistake was made in the formulation of the original law. That mistake has never been corrected – until now. When Eddington and Einstein warned scientists to stay away from tinkering with the second law, including entropy, scientists paid attention—for the last 150 years.

Our new research shows the hidden and deepest force underlying everything in nature is actually a *creative* force that brings about more, new and deeper relationships, connections, among all the elements in the universe—from subatomic particles to humans and human societies.

Entropy is the reverse of what science has taught and thought. This underlying hidden creative force, working in the background, affects everything. It is why we have such powerful technologies, why we have beautiful flowers, why we live longer. Why we pay so much attention when things go wrong. And why our world is absolutely destined to get better—much better.

And why the most basic and beautiful beliefs of spirituality and religion are in total *agreement* with this new scientific discovery!

This force has been hidden so well that we simply have not been consciously aware that this force can be made available to deliberately enhance our lives. Some people have known it and used it intuitively and instinctively but could not explain to us why their lives were working so well, though in many cases they have tried and even been willing to accept being called "loony" for trying. We are very happy to help them out—and make this information available to everyone. This book is not an attempt to sell you "quantum" water or "Superforce workshops" or other kinds of fruitloopery.

We are presenting a new scientific law as a well-substantiated explanation of the most fundamental behaviors of the natural world, based on a vast body of facts that have been repeatedly confirmed through observations that range from subatomic phenomena to human societies. This is a fact-supported concept—not some speculation, but a theory based on an enormous array of reliable and testable accounts of the real world.

We can promise you that you will find in this book not only the scientific evidence that supports these claims, but evidence that also is in agreement with what we have held in our hearts about our deepest spiritual beliefs. Unveiling new awareness, understanding and conscious use of this vital force in nature will assist us to enjoy even better human relationships. It will help us tap into our inborn creative capabilities to aid us in generating the kind of positive future we desire.

This universal force can operate on our behalf once we know it is there and how it works.

## IN THE BEGINNING

Our search began in such an odd way that we would like to take a few minutes to share with you how this book and all the accompanying research began. That we would overturn the most basic law of physics and open the door to a new, positive and creative relationship with the divine was probably the most remote idea in our minds when we started this book.

This book started a number of years ago in an extremely simple and straightforward way. After over half a century of following our passion – our obsession if you will – with the phenomenon of creativity and transformational change, we had two goals in mind.

The first was intended as a follow-up to our two previous books on creative change. The first by George Land was *Grow or Die, the Unifying Principle of Transformation* (Random House, 1973). It showed the links among the various levels of nature that demonstrated three different kinds of growth, change and creativity. Although these had been identified in different disciplines with different words, they actually worked the same repetitive way in simple natural patterns and could be seen in human activities, as well. This was followed by George Land and Beth Jarman in *Breakpoint and Beyond, Mastering the Future—Today* (HarperBusiness, 1993), which showed how the principles of transformational and creative change applied in organizations and even societies.

Over the past ten years, having done a great deal more research and having worked and done investigations with almost a hundred organizations around the world, we decided that our experience and new findings required a new book. So our first goal looked uncomplicated—just a follow-up to our prior work.

The second goal was more ambitious and exciting: to offer

something unique. We would strive to answer a question that had always beset us: what was it that drove this process of creativity in nature, in evolution, and in humans, sometimes against all odds? Beth, in particular, as a top executive in a number of large organizations, had experienced the joy of seeing her colleagues, while facing meager rewards and sometimes considerable opposition, express remarkable creativity.

So what was the root that spawned this flourishing of creativity? We had never seen an answer to this question emerge from all of the centuries of research and explorations of science. The "why" behind creativity was totally obscure.

We had better define what we mean by our use of the word "creative" or "creativity" in this book. Basically creativity occurs when two or more "things" (materials, designs, words, concepts, ideas, and movements – as in dance) are connected and bring about something that is new, novel, and different from the past. It is more than the sum of its parts. Creativity is unpredictable, more complex, and connects well with the environment.

Different materials are put together and we have a light bulb. A painter mixes colors and applies them to canvas and we have a painting. Mom tries a new mixture and ends up with Mexican lasagna. Five-year-old Timmy takes a box and makes a spaceship. These are all "creative." Creativity includes a vast array of activities.

We became more excited in our search as a number of scientists had begun to re-explore something that is called "emergence." This term is used to describe physical events in nature lying deeper than human phenomena: things like atoms, molecules, even stars and galaxies. These events are material phenomena that meet the exact descriptors that can be applied to human creativity. Included are such events as the very first atoms, the first complex molecules, the first life, and the appearance of consciousness. Although no scientific explanations have been given as to why these things could or should happen, at least science recognizes that such amazing events do happen – and with regularity, from the Big Bang to evolution.

Creative acts have a giant range but all share the same attributes—novel, new combinations, more than the sum of their parts, unpredictable, different from the past, more complex and connect well with their environment. Forgive us for repeating. The idea of the creative or emergent is just too often misunderstood. In this book the definition is absolutely critical.

These recognitions of creativity in natural science forcefully spurred us on in our search for what might be the natural force driving creativity both in nature and in humans.

One powerful example of this new discovery about the positive force in entropy is that it revealed a hidden creative potential that lay at the foundation of some major positive changes people have made in their lives.

We have all been exposed to the sort of "personal miracles" people have experienced that have been reported in many books such as *The Power of Positive Thinking* and the like. They have extolled the idea that if we change the way we think, wonderful things can happen. We just envision our goal, our desire, or our dream, keep that in mind, and it will somehow come true. And sometimes, for some people, it does. Yet the authors of these books do not tell us *why* these good things happen. They just asked us to believe that if you think it, it will happen.

Both of your authors have read many of these books. But somehow we were not moved to change. We realized that something was lacking. We couldn't help but ask what was the reason why these ideas should work? The concepts seemed simplistic and naïve and not supported by any rational foundation. We remained unconvinced, so we did not take advan-

tage of what we found later to be intuitive wisdom. Einstein once said, "The intuitive mind is a sacred gift and the rational mind is a faithful servant. We have created a society that honors the servant and has forgotten the gift."[7]

We found that after peeling back layer after layer of reality and scientific discoveries, the hidden force was revealed and clearly showed that there is, after all, a real natural energy operating behind these concepts of "positive thinking" or "the law of attraction" or "miracle thinking" or whatever the latest thinking on the projection of future success happens to be. In our search for the roots of creativity we ended up also wanting to know why and how this concept of thinking might work in the "real" world.

With the discovery of another hidden force, what has come to be known as "quantum physics," and a new understanding of the law of entropy we discovered, a previously unknown real, physical force has been exposed. The result is a real natural law, moving behind the scenes, that contributes to making such thinking work. This law reveals that a deep force is behind what makes, for example, subatomic particles merge to make atoms, and atoms get together to form molecules, and molecules join to make life. It causes all the other miracles of nature's own "positive thinking" and "law of attraction."

This book is about a scientific discovery. But it is really about each of *us*. It is about understanding the reason, the reality and solid foundation supporting how our thinking affects our physical and spiritual lives and our future together.

We do not want to practice some kind of sham humility here, but we do not want to take real credit for what this book

---

7. Quoted in Jens Haase and Egil Boisen, "Neurosurgical training: more hours needed or a new learning culture?", *Surgical Neurology*, Volume 72, Issue 1, July 2009, 89–95.

contains. Surely we have dedicated decades of hard work to it, but the facts are that we actually put together a jigsaw puzzle made up of the enormous efforts and discoveries of many others who worked much harder than we, over centuries, peering into the inner workings of nature. Why has it taken so long to bring this puzzle together to find this new force? For one thing, of course, were the warnings of great scientists not to try to change the old law. Who were we to challenge a law that had been unquestioned for a century and a half by the entire scientific community? For another it took bringing together findings from different disciplines and this is difficult in a world of specialization and barriers. Another reason was simply that new discoveries had not brought about a reexamination of the old law.

To uncover and understand this wondrous force, this book will *not* need to be about some new mathematics. Whew! It will be about the real nature of this deep creative force that guides our universe and our lives. It can be understood in everyday language. To understand the simple beauty of nature does not require abstruse language or math. As it turns out, we discovered that nature actually works in pretty simple but quite surprising ways.

The vast accumulation of facts about natural emergence has recently been expressed in such books as *The Emergence of Everything: How the World Became Complex* by Harold J. Morowitz (Oxford University Press, 2002) and *The Re-Emergence of Emergence: The Emergentist Hypothesis from Science to Religion* by Philip Clayton and Paul Davies (Oxford University Press, 2008). All emphasize the enduring absence of any explanation or understanding of why or how these amazing events occur at all.

Our jigsaw puzzle, made up of these pieces and the abundance of other new data, finally revealed the taproot of creativity that feeds the tree, branches and fruits of both the ordinary, and the exceptional, emergent and creative acts of nature and our lives: those that whither and those that flourish.

The central concept and explanation that will emerge in these pages was such a surprise to us, and probably will be to you as well. It meets the great goal stated by Albert Einstein: "I have deep faith that the principle of the universe will be beautiful and simple."[8]

We found this to be so.

In this book we will share with you how we dug down in science and found a remarkable fact that translated into profound meaning. Where does creativity come from? What does that mean in my life? What could that mean as we create our new and different future together?

Our discovery has overturned that seemingly inviolate fundamental scientific law of entropy that has heretofore denied us a sense of meaning for ourselves and even for the universe itself.

Amazingly, the unearthing of this hidden force behind entropy has unified the most fundamental law of science with the shared spiritual concept of a universal and ever-present creative God!

We thought, at first, that our very deep longing to find such a positive and holy force in the cosmos had clouded our vision in such a way as to distort our perception of reality. We know the brain has a tendency to trick us this way. We see what we want to see. As Einstein also put it, "The theory determines what we are allowed to see."[9] We have done our very best to overcome this natural proclivity. In these pages you will find abundant hard scientific proof to support what we have found.

---

8. Quoted in Stuart Kauffman, *At Home in the Universe* (Oxford University Press, 1995).

9. Edmund Blair Bolles, *Einstein Defiant: Genius versus Genius in the Quantum Revolution* (Washington, DC: Joseph Henry Press, 2004).

We have also avoided using metaphysical or esoteric language or abstruse scientific equations. As we have said, you will find this book written in everyday language. It does not require some level of "genius," or enormous formal education, to understand and apply or produce ever more human "emergences" and creations to better the human condition and our everyday lives.

The idea is that not only is this scientific discovery solid, but unlike many current views of natural laws, any one of us can use plain common sense to decide if this breakthrough truly "makes sense" and gives new and deep value to our blessed daily lives and in our big picture of the planet and the universe we inhabit.

Our search started again in a totally unexpected way when we were visiting Northern Ireland on a trip for a client. We had an odd experience that completely reset the direction of our expedition in this book. We were pointed to the fundamental natural law that dramatically cancelled out any idea of finding a natural force that could drive creativity. In the end, what affected us most profoundly about this discovery was how it can affect the way we think about our world—and ourselves.

CHAPTER TWO

# NORTHERN IRELAND

---

## NORTHERN IRELAND, 2001

It was dark inside, just as it was outside, dark clouds moving toward us, erasing the horizon line on the ocean across the road. We had stopped at the pub on the coast of Northern Ireland on our way to see an ancient castle. Our guide, Jerry, insisted we have an authentic Irish stew before we headed home. We had no clue when we turned into the pub's parking lot that our lives would also take a turn that day—a momentous one.

The pub's sign indicated it had been one of the first in the country and we didn't doubt it, heavy as it was with the almost black walls stained with smoke and beer fumes. "You're from America are ye?" and the conversation quickly revealed cousins in the New World. A cheery bunch they were, seven gents from the village welcoming us to their hangout.

In no time we were immersed in the local humor, prompting us to buy a round for the house. All accepted, save for a young man buried in a big book at the end of the bar. He was dour indeed as he shook his head, turning our offer down. "Oh, and why not Pad?" the bartender asked. "You look like somebody died, lad." "It's worse than that," the young man responded. "Oh and has the sky fallen?" one of the men said. "That's the least of it," said Paddy. "So, share it with us lad and a pint too?" said the bartender as he passed an ancient rag across the bar. "You look like you could use a little cheer."

"Look," responded Paddy, "you know I'm up at college studying up so I can get into university and I'm trying to get ready for the big test." He pointed down at his book. "I just don't know why I'm trying at all." "Is it too hard for your dim brain?" one of the men joked. "Not so," said Paddy. "Here, just look at this." He passed the book over. The fellow he handed the book to read for a minute, shook his head, and looked up.

"What's this nonsense?" he asked. "It's the latest book in physics," Paddy answered.

"Listen to this," said the man, reading aloud from the book. "One of the most important and fundamental laws of science is the law of entropy," he continued. "This is what it says here, 'Entropy is a universal force which causes everything in nature to inevitably disintegrate into lower and lower levels of organization. The world, in essence, is a great machine running down and wearing out. No matter what we do everything will end up in cold chaos.'"

"Well, that's a great thing to know on a gloomy day like this. Thanks lad for cheering us all up." "Say that again," said Mike, the bartender, with a querulous look. And then he passed the book around so everybody could read it. "You see," said Paddy, "what's the use? Why should I go to university where I can learn more about how the world's going to hell?"

"Well," said Mike, "that calls for another pint, wouldn't it?"

With the lull in the conversation we asked the bartender for three orders of their famous Irish stew and moved over to a table. As we sat down we noticed that Paddy had shifted and was looking at us. We gestured at him to join us and soon he made a fourth at our table. "I'm George and this is Beth and Jerry." "And I'm Paddy," he said. "What'll you have?" the bartender asked. "Just a wee bun," answered Paddy. He seemed very agitated. "What I showed those fellows was the least of it!" Paddy declared. "The rest of it makes me away in my head! Wait till I tell ye."

"What do you mean?" we echoed. "That's the bleedn' problem," said Paddy. "You probably are wonderin' why I'm so old to be goin' to college? Well, I was the one in my family picked to be a priest. I went to seminary. But I just didn't have a bleedn' *faith*. So, I finally knew science was the answer for me. Boggin' science! Wait a minute." He reached over to his bar stool and dragged his backpack over to the table. He opened it up and rummaged through it until he found a tattered notebook. Laying it on the table, he paged through it until he found a page complete with big X's and exclamation marks.

"We've been lookin' at videos by big Nobel Prize scientists and here's what the top cosmologist says: 'I have been studying the universe for over fifty years and what I have found is that the universe has become ever more complicated and ever more meaningless—*meaningless*.'" He turned a few pages. "The fellow that followed him, who has written about a dozen books on science said, "'Try as they might scientists have never found even a whiff of purpose to the universe.'" Paddy looked up, sadness in his eyes. "It's all just a crap shoot, random chance. Oh, and by the way the top physicist in the world, Richard Feynman, says 'Nobody understands quantum mechanics.' The universe is about as useful as a bottle of crisps!" He paused for a minute, lost in thought. Finally, he said, "What kind of God makes a world like this? It's going to hell and nobody can understand it. I can't be a priest or a scientist."

By then our stew was on the table and Paddy looked up at the big wooden clock over the bar. "Me mom will have somethin' on the table. I gotta go. As they say on the American telly, life sucks! Cheerio." With that, he grabbed his backpack and was quickly out the door. He hadn't touched his "wee bun."

We sat there pretty much in shock.

That night, as we packed and got ready for our trip home, we didn't talk much, a very unusual situation. It wasn't until the next day, after we had flown from Belfast to London, made our way through the labyrinth of Heathrow, found a great bookstore, had breakfast and settled into our "upper class" seats (courtesy of our generous clients), that we began to share our thoughts and feelings.

A long and unsettling conversation spilled out. A lot of it seemed to have been waiting a long time to be said. It was as if Paddy had been our own son, and somehow represented all humankind. "How many Paddys are out there?" "I hear Paddy every night on the news. There are many people who just seem lost – who don't seem to care about anything." "And even when you get some good news there is always a big BUT..." "I remember that gang leader who looked right in the camera and said, 'Life sucks.'"

It went on like this. Our own growing negativism and pessimism and depression was just not us! Over four decades, we had always kept a high level of excitement and enthusiasm working with people, many times ordinary workers from around the world, helping them get in contact with their own creativity, the wonderful skill that had been suppressed ever since they started formal schooling. The lights in their eyes, their great enthusiasm when they discovered their real inborn talents could be rekindled with just a few hours of encouragement. This has been a great joy in our work. We have experienced the elusive "human potential" awaken so many times and in people no one expected would ever have a creative idea. We truly knew the creativity of the human spirit, as they say, up close and personal.

George (a book nut) had picked up a book at Heathrow in the Oxford publications section. "You won't believe what I found," he said. "This fellow that Paddy mentioned has written *seventy* books on science," and he proceeded to read from the book he had found.[10] "On page eight, 'regarding the findings of science, what does it reveal? Sniff as it might, it (sci-

10. Peter Atkins, *On Being* (Oxford University Press, 2011).

ence) finds not the slightest hint of purpose in any event it has examined.' It goes on, 'In broad terms, the second law asserts that things get worse. A bit more specifically, it acknowledges that matter and energy tend to disperse in disorder.' And 'that's all there is to natural change: spreading in disorder.'"

"That's just not true," said Beth. George nodded.

We did not realize it until weeks later that our lives had taken a new direction. We had abandoned all of our other work and started our search for an understanding of what all of this really meant. We deeply believed that something could explain what we had experienced in our lives: something that violated these "laws" of nature — something that we and the Paddys of the world could believe in.

We were determined to learn much more about "entropy," this potential dead end of our exploration. It ended up provoking entirely new questions. As we went along, our inquisitiveness grew more and more intense. Treading an unbroken path, we found ourselves in some very unusual territory. It took us into the notebooks of scientists who lived in the 1800s and the correspondence of scientists like Albert Einstein. Like being sucked into a whirlpool of unending questions, we pursued, of all things, assumptions about how gases interact at the atomic level, the causality of special relativity versus quantum mechanics and the neurobiology of how we think about the future.

We reached a point in unexplored territory where we actually became very alarmed about the completely new direction that was literally taking us over, carrying us into a new and, frankly, intimidating world. Did we have the skills, the knowledge, or even the audacity to continue to follow this nerve-racking trail to the source of creativity, if it existed at all? It was looking more and more like it might be the enigmatic source of meaning itself. By "meaning" we mean "Why are we here? What is our purpose?"

The reason for our book is to share with you a discovery we made that transformed our lives and might do that for you as well. When we finally found what we had been looking for, we were staggered by its implications, its simplicity, its beauty, and the many other questions it answered. We ask you to share in our journey and understand how we arrived at our conclusions and how to apply what we have discovered in your own lives.

Our diverse backgrounds contributed to what we found. George is basically a scientist: what is called a "general systems scientist." This is the study of uniting many different fields like physics, chemistry and biology to discover the common laws that affect them all. Beth has been devoted to understanding human potentials and how they might best be realized. In both cases the question of creativity has been central: how nature expresses creativity through continually evolving and bringing about entirely new things from subatomic particles to the dynamics of human thinking and relationships.

Finally, we believe we can make sense of the universe and nature and our place in it. We continue to be shocked that this profound mystery can be uncovered—and that it could contain such beauty, elegance and clarity, and could have such influence in our personal lives!

At first, we were so stunned that we were almost incoherent. We actually questioned our own reasoning. It was like, as Thoreau said, "Finding a trout in the milk." Had our conviction that such a groundbreaking idea could be discovered at all led us into believing something unreal? Had we merely put on some rose-colored glasses that allowed us to see only what we wanted to see? We know this has happened to many people. As, little by little, the surprising concept we had found could be proven and documented by hard science, ironclad facts emerged one by one that showed we were on the right track. At last, we felt we had gained enough understanding and evidence to share what we found even though it violated what science had believed for a century and a half.

Although what is happening around us in today's world seems chaotic, the fact we now know is that nature is transforming the planet in very positive ways. If we know what is going on we can be woven into what she is trying to do. How can we use nature's blueprint for creative change?

In these pages we will find that nature reveals a deeper understanding of the powerful forces guiding and shaping the universe. It will reveal the inner knowledge that each of us holds.

We are humbled by our discoveries. We offer them to the reader with the overwhelming desire that together we will better understand nature's true driving forces: those that move us forward together and help us find new solutions to our most intractable problems.

CHAPTER THREE

# GETTING TO THE BOTTOM OF THINGS

When we returned from Northern Ireland we were feeling very unsettled. After doing so much research over the years on creativity, the conversation in the pub and the book George found in the airport had suddenly and surprisingly focused our attention on something in science we had never before considered—*entropy*, part of the second law of thermodynamics. If you are like us, you may have heard about this seemingly esoteric law in a classroom somewhere or run into it in a popular book of the same name, but it surely is not a term in popular usage. Yet, for our purpose (trying to find the source driving the power of creativity), we realized entropy had very fundamental meaning. It is a basic law that says how nature behaves everywhere. The law of entropy is used to explain the activities of the smallest atom to the most gigantic star and galaxy—and everything in between. So naturally we would look there to find clues about creativity: for example, how all the universe's atoms, stars, etc. were "created."

Before we get into any details about entropy, we have to take a bit of a digression about the very first stumbling block we encountered in our attempt to understand this powerful law that guides everything in the universe.

As we first began to get the facts about entropy we ran into some mathematical expressions we did not understand. Neither of us is a mathematician so we called a friend who is one and who teaches advanced mathematics at a New York university. In the course of our conversation he shared a fact with us that we had never considered. Since entropy describes a law about disorder or destruction if you will, we also asked about scientific mathematical expressions or equations relating to creativity – or what we noted that science calls "*emergence*."

If you are like us, you might be surprised at the answer.

"There is no such thing," he said.

The revelation that scientific thinking contained a giant built-in obstacle to the search for creativity itself was a shock. Rules that were set up by science centuries ago contained limits that remain pretty much unquestioned.

Ever since Galileo Galilei proclaimed, "Mathematics is the language with which God has written the universe," we have had to face the fact that our science has been confined within a mathematical prison of thought and dogma regarding our deepest nature and the truth of our connection with the universe.

We found that this problem with mathematics contributes greatly to creative processes slipping through the net of scientific investigations. It has missed the underlying forces driving the cosmos.

The idea that our very well-developed sciences could actually have overlooked something crucially important is a very bold statement. Before you read on consider this one critical fact: Mathematics is the absolute basis of all of our physical science and technology and of much of our thinking and conclusions about the universe.

We depend on it to inform us, from the working of the cosmos to interpreting the scans of our brains.

Around 400 BCE the philosopher Plato stated, "Mathematics is the language in which the gods speak to people."

The idea held on. In the thirteenth century Roger Bacon said, "There are four great sciences ... Of these sciences the gate and key is mathematics, which the saints discovered at the beginning of the world."

Later, Einstein noted, "What we call physics comprises that group of natural sciences which base their concepts on measurements; and whose concepts and propositions lend themselves to mathematical formulation. The realm is accordingly defined as that part of the sum total of our knowledge which is capable of being expressed in mathematical terms."

William Byers, professor emeritus in mathematics and statistics at Concordia University, concluded: "Mathematics today is obsessed with rigor, and this actually suppresses creativity." The emphasis on rigor and technique has led to a "lawyer's vision of math, where the main goal is the nit-picking avoidance of mistakes."[11]

Yet, as we look at all the most vital natural happenings in the universe, the very most important events are those which produced new, different, unpredictable and more complex phenomena. They include happenings that range from the appearance of the first atom to the first complex molecule, to the amazing advent of life itself, and ultimately to the consciousness that can ask why any of this happened in the first place. In human terms we call such novel happenings "creative," as they have the same properties as nature's "emergences," but come from another recent emergence—our minds.

No mathematics, scientific formula or concept has ever been found to express, explain or understand these emergent or creative happenings.

Mathematics can deal with quantity but has great difficulty dealing with changes in *quality*, particularly qualitative change.

According to the famous mathematician and philosopher

11. William Byers, "Less proof, more truth: Review of *How Mathematicians Think*.".

Alfred North Whitehead, "Classical science, which originated in the seventeenth century, was an example of misplaced concreteness unable to express creativity as the basic property of nature."[12]

The award-winning mathematician László Lovász commented on his encounters with the complexity in which the parts of a structure are completely determined (and thus can be dealt with by math), and how when put together they result in new behaviors. He said, "You need entirely new notions, and a new phenomenology to describe the behavior of the structure...(for one thing just to write up and solve the equations would be impossible, and I think this impossibility is very serious in any sense)."[13]

The foundational stone of science is that any hypothesis or idea must be proven by "repeatable" experiments. In the case of emergent or creative phenomena, how can anyone expect that novel phenomena can be replicated? Emergent and creative phenomena can be by their very nature impossible to replicate. Emergent phenomena very often cannot meet the traditional repeatability test of science.

In light of the modern discoveries of science like quantum mechanics, the renowned scientist Richard Feynman said about these old scientific rules, "It is necessary for the very existence of science that the same conditions always produce the same result. Well," Feynman continued, "they don't. You set up the circumstance here and the same conditions every time, and you cannot predict behind which hole you'll see the electron. Yet science goes on in spite of it—although the same conditions don't produce the same results. That makes us unhappy, that we cannot predict exactly what will happen. Incidentally,

12. Quoted in Ilya Prigogine, *The End of Certainty* (New York: Free Press, 1997), p.59.

13. *The Emergence of Complexity,* Proceedings Plenary Session of the Pontifical Academy of Sciences, October 27-31, 1992, p. 80.

you could make a circumstance which is very dangerous and serious, and man must know, and still can't predict."[14]

Feynman is describing what we can now see as a kind of hinge point in history – the history of thought itself opening a new door to truth.

Flowing from the original discoveries of the Arabs and Greeks, a particular way of thinking emerged in civilization as *the* way to think about our world. Aristotle's logic was further developed by the philosophers in the mid-fourteenth century. The development of modern mathematical logic during this era is the most significant in the history of logic, and is one of the most significant and outstanding events in human intellectual history – particularly as we attempt to understand the real creative workings of nature, and how we could have gone so wrong in our conclusions about how the universe works.

Without anyone noticing, the concept of mechanical "reason" and logical thinking became the accepted basis of human thought. As we will see, this form of thinking denied us a true understanding of nature.

It cannot be emphasized enough that logic and reason, what became the scientific method, became the standard operating method adopted for human thinking. Other platforms such as those incorporating creative or imaginative thinking were left behind as civilization moved ahead. Newton ruled.

This is why it became such an astonishing surprise to discover through quantum mechanics, finding random behavior throughout nature, that at its most fundamental level, nature does *not* follow the rules of logic or reason. This was inconceivable thinking. We had expected the universe to be

---

14. Richard Feynman, *The Character of Physical Law* (Modern Library, 1994), p.141.

organized in a logical, orderly fashion and to obey uniform laws because the universe was supposed to have been created by the power of a mechanical, reasonable God.

Logic deals with fixed and unchanging categories. These are good enough as approximations, but now we know they simply do not reflect reality. Science has succeeded in ignoring prickly irregularities of nature. The very emergent, creative and unpredictable phenomena that are the universe's most important and fundamental drivers did not fit our ancient and accepted paradigm of logic. The universe as we know it (and we ourselves) would not be here if these incredibly unreasonable, creative, unexplainable, unpredictable "accidents" outside of logic had not occurred that somehow, "accident after accident," produced intelligent, creative, and feeling beings. For all of this to come about from chaotic stardust, it seems that there must be a deep force underneath driving it all.

There is no math or repeatability or standard thinking to call on to help our search. We will need to call on non-standard thinking to grasp the beauty, simplicity, hope and promise given us by nature.

And the law of entropy – as it is defined today with today's limited thinking – is the *opposite* of creativity. The limits of logical thinking ended up confining us to a prison of thinking that produced a law of destruction, not creation, to explain the universe.

When we first began looking at information about entropy we found the definition – from many sources:

One recent definition we found comes from the *Oxford Dictionary of Science* (2005): "Entropy – a measure of the unavailability of a system's energy to do work; also a measure of disorder; *the higher the entropy the greater the disorder.*" (Emphasis ours.) We have included a number of other definitions of entropy in Appendix Two.

In the 1850s the physicist R. J. E. Clausius found that "The total entropy change always increases for every naturally occurring event that could then be observed. Thus, disorder is continually increasing everywhere throughout the universe."

Let's review, realizing that entropy is a universal "force" that causes organized forms to gradually disintegrate into lower and lower levels of organization. Clausius and the ultimate definer of entropy, Ludwig Boltzmann, saw the universe as a great mechanism wearing down and never going in the other direction.

Research on the concept of the second law of thermodynamics (the movement of heat) began with the efforts of Carnot and Clausius but it was left to Ludwig Boltzmann in the mid-1800s to pull everything together into a comprehensive set of scientific laws in his *Lectures on Gas Theory,* a 490-page tome on the behavior of gases. All of the laws rest on the foundation that heat flows spontaneously from hot to cold bodies, but the opposite never occurs, and *increasing disorder always accompanies this flow.*

Thus, along with heat movement, the conclusion reached was that our lives, nature and the universe are in a state where disorder will increase everywhere. By disorder we mean unpredictability, chaos, and confusion.

We want to quote again Jeremy Rifkin's book *Entropy,*[15] which says, "In essence, the second law says that everything in the entire universe began with structure and value and is irrevocably moving in the direction of random chaos and waste." This concept percolates around the base of Murphy's Law, "If anything can go wrong, it will." This is often cited as a form of the second law of thermodynamics because both

15. Jeremy Rifkin, *Entropy: A New World View* (New York: Viking Press, 1980), p. 6.

are predicting an inevitable predisposition or tendency to a more disorganized state.

The more we found out about the second law of thermodynamics and entropy, the more discouraged we became about finding a driving force for the creative process. Entropy decreed that at the taproot of all natural processes science has found —without exception—destruction, not creation. How could that be?

Well, obviously this law is important to us, as we are looking for some force that could be the source driving the process of creativity. But what might it mean to everyone else? Since Newton's time science has become the way to know about our world, the way to understand our world—as best we can—with knowledge and its accompanying technology. We have looked to science to answer the big questions: where did we come from, what are we made of, how do things work, where are we going, and a host of other human curiosities. Yet science has avoided trying to deal with the really big questions, like "What does it all mean?" Science might avoid such questions because its most fundamental law points to no meaning, or rather to ultimate chaos, whatever that might mean.

How can we get excited about that kind of future? What kind of higher power would create such a design? With no fundamental scientific law to turn to, where do we look for anything driving creativity?

Since the 1900s two gigantic revolutions uncovering hidden forces have upended the world of physics: the theories of relativity and quantum mechanics.[16] Yet since the 1860s the law of entropy has held fast, as holy writ carved in stone, without the least modification.

16. The theories of relativity are the theory of special relativity and the theory of general relativity."

We must emphasize and be extra clear about this law. It comes in *two* parts. One part is absolutely correct! That part, identified early on by Clausius, simply shows that heat always moves from hotter areas to colder areas. Upon investigation, equations were developed to exactly predict this movement and this has been a great boon to anyone working with engines. At the time it was quite a revolution. The part about "disorder" and entropy was just added on and has become accepted along with the movement of heat.

This may sound rather dull, but the movement of gas and heat was the "big thing" of its time. It solved the biggest and most important problem occupying scientists, engineers, industrialists, investors and even ordinary people at the time. It was the era of the Industrial Revolution and the big question was how to get steam engines to work correctly and efficiently. And the second law of thermodynamics (including entropy) solved it. Roaring into every person's life were railroads, factories, great ships, and giant machines, all geared to take the toil and sweat out of humanity's struggle for existence. It was a very big deal indeed!

In Boltzmann's book, some 275 pages were devoted to the movement of gases. About five pages at the end mentioned the implied fate of the universe, moving inexorably toward disorder. The ensuing big news about steam engines completely overshadowed his ideas about the ultimate fate of the universe. It was like a faint shadow following the bright comet of steam power. Who cared about some far distant possibility when the immediacy of the miracle of the steam engine was right now! So the disorder law was tacked onto the second law and hung on for the next 150 years without serious reexamination.

As we look out of our study window at our backyard, we are stunned by the beauty of nature: the grasses and flowers, the trees and fruits. It is magical and munificent – and we live in the desert. The truth of nature, the miracles, and the fact that we are here to wonder at it at all, is a vast puzzlement to science. Science accepts a basic law that this magic of nature should not be happening. The entropy law says that everything inevitably gets worse; it shows that matter and energy tend to spread in disorder, and that there are many more ways to be disorderly than to be orderly.

This book will show that the science of that era got it wrong – that the beauty and creative magic of our lives is *the* reality. Once understood, we can tap into this creative process to make everything even better—much, much better.

Nature has also shown the way to move forward in a creative way, showing how to join disparate elements to produce new and different possibilities. We can learn this natural method of recreating our planet to benefit everyone. Nature continually transcends the past and converts what are often seen as problems into amazing opportunities for progress. We can do this too!

Today's chaotic world is like being at sea with the wind whipping up waves into a frenzy in one direction, but ignoring that beneath the chaotic waves lies a much more powerful hidden force, carrying us in a totally different direction.

We were greatly encouraged by the distinguished Nobel Prize winning scientist, Albert Szent-Györgyi. He proposed in the 1970s that there *must* be some "innate force" in all living things which acts to order things at higher levels. Although he never found the solution, he proposed the word "syntropy" to refer to this innate force.[17]

In the winter of 1927 Alfred Kerr, at a dinner party in the home of the publisher Samuel Fischer, asked Einstein about his religion. Einstein replied, "Try and penetrate with our

17. Albert Szent-Györgyi, "Drive in Living Matter to Perfect Itself," *Synthesis* 1, No. 1 (1977) p. 14-26.

limited means the secrets of nature and you will find that, behind all the discernible concatenations, there remains something subtle, intangible and inexplicable. Veneration for this force beyond anything that we can comprehend is my religion. *To that extent I am, in point of fact, religious.*"[18] (Emphasis ours.)

Notwithstanding the suggestions by these eminent scientists, little attention has been paid to looking for this hidden force that might foretell a very different fate for us and our universe.

We are going to dig very deeply to discover the foundation of a revolutionary new view of nature that utterly reassesses our place in the universe.

Why are we here, what is the purpose of the universe, of life itself, where are we going? As one group of scientists we were working with put it when we mentioned our quest for this elusive force, "Let's not go there!" These are not even considered proper questions for science. Science is focused on and has been superbly successful in answering the "how" of nature; it leaves the "why" to philosophy and religion. Here we will be crossing that line, stepping into the *terra incognita* between, if you will, spirituality and science.

We believe it is useful to understand how this very fundamental "law" of science came into being and why people believe it. Science has accepted the "second law of thermodynamics" and the part known as "entropy" as correct for over a century and a half. If we would have stated that this law is one of the most important things we were going to examine in this book, we suspect most readers would have put the book aside and would have found something else to do – almost anything else. We wouldn't have blamed you. Who

---

18. Quoted in H. G. Kessler, *The Diary of a Cosmopolitan* (London: Weidenfeld and Nicolson, 1971) p.322.

cares about this law? It is taught in every science classroom on the planet, but quickly forgotten—or perhaps unconsciously denied. Could the universe really be designed this way? Certainly no religion would agree with this "supreme" law!

It may seem that we are beating a dead horse here, endlessly repeating ourselves. And perhaps this all seems boring as well. But the law of entropy has been insidious, subtle and even dangerous. It has worked its way into becoming an essential basis for every science, from physics to sociology and psychology, and has achieved legendary status as a fundamental law of natural science. And we don't like to talk about it. Freud has taught us that since we are always comforting ourselves and avoiding facets of our experience we would rather not face, our ideas tend to be rationalizations by which we put a good face on a bad thing.

Why on earth would you want to read a book about a law of nature that points out something at some level you already know—and would just as soon not like to be reminded of? And this so-called "heat death" is a long way off. So things do wear out. If you don't maintain your car it falls apart; likewise, your house. Everything! Just look around: failing businesses, failing banks, failing schools, governments that don't work. As we often hear, "Things are going to hell in a hand basket." All of this, we are told, verifies that as a law of nature, entropy is working. It's pretty discouraging. Science tells us that in the end no matter what we do disorder and chaos will always catch up with us. What a depressing and insidious idea!

If all of this were not enough, science also points out that this activity of nature is *spontaneous* and *irreversible*. We don't have to make our car disintegrate, it does it all on its own. It's just natural. Things just get disordered without anything to help them. It's spontaneous and it goes only in one direction. An "arrow of time" pervades nature. The law assures us that disorder doesn't ever move backwards into order.

Entropy is a sad and depressing concept.

Yet Boltzmann was right—when it comes to *heat*. This basic concept and law does apply to heat. What we can now see, regarding the second part of Boltzmann's law, is that the gas itself and how it behaves is hidden and totally misunderstood.

Sure, your car or house breaks down. But another, even more powerful law is working in the background. It is the law we call *Creative Connecting*. Billions of years ago this law *spontaneously* and *irreversibly* took inert rocks on our planet and transformed them into all the living things around us and, more recently, into us. Rather than disorder, this law, when understood in a new way, predicts that our world is not "coming apart" but moving inevitably to ever more interconnected order. How can we explain this when entropy is the accepted law?

An enormous mistake was made in the interpretation of the natural force of entropy.

This book will show that *exactly the opposite of disordering is happening!* This book provides a new understanding of nature's laws of order. It makes for optimism and excitement about what's to come in our world and, more than that, shows how each of us can take advantage of nature's amazing Creative Connecting power.

As we begin to grasp the concept of what we call "Creative Connecting," we are starting to catch a glimmer of a profound purpose behind the activities of the universe.

A number of philosophers, particularly Tilliard de Chardin, proposed that in the life process, we continually find a decrease of entropy occurring. Living organisms experience increased organization, both in their own world and also in what organisms create in the world around them, reducing disorder and, in essence, somehow opposing the universal law of entropy by creating an ordering process.

We can see from a scientific basis that this "ordering" process even goes much deeper than life; it also extends into the so-called "non-living" cosmos.

This idea of increasing ordering and the term "synergy" was used by Luigi Fantappie in his prescient book in 1941.[19] Szent-Györgyi, Fantappie and Chardin's ideas of a positive side of entropy were dismissed by the scientific community.

The second law was too ingrained and universally accepted.

The business of re-examining a basic scientific law in a new way could certainly be off-putting at the very least. It smacks of complicated mathematics and equations of the kind we perhaps shuddered at when we were students in science class. This is not the case here, as you will see. Not only is it not complicated, it is something that is actually intuitive and quite easy to grasp.

The noted physicist Stephen Hawking stated, "If we do discover a complete theory, it should be in time understandable in broad principle by everyone. Then we shall all, philosophers, scientists, and just ordinary people be able to take part in the discussion of why we and the universe exist."

Most importantly, a new understanding of this very fundamental law of nature makes up a solid foundation for dealing with the vital questions we humans have had on our minds since we began to question our place in the universe and the meaning of life. It allows us to see and understand the nature of the universe and our place in it in a new, confident and hopeful way. When someone says "Get real!" we have a new and positive answer we can believe in.

Let's look at one of the most basic explanations and experiments that "prove" how the universe treats matter in

---

19. Luigi Fantappie, *Principi di una teoria unitaria del mondo fisico e biologico* (Rome: Di Renzo Editore, 1993).

the Second Law. The following is the thought experiment done by Ludwig Boltzmann and many others to explain and prove the Second Law.

A "thought experiment" is conducted purely in one's mind when there is no way with actual existing experimental equipment that it can be conducted in a laboratory. This is often done in scientific research. Galileo described his thought experiment about dropping stones. Albert Einstein was famous for his thought experiments that revealed his discoveries of relativity. In his case he imagined riding a beam of light to find out how light worked. The result gave him the insight to create the theory of relativity.

So let's look at Boltzmann's thought experiment and proof of his concept of disordering.

Imagine a box divided in half by a wall. You introduce a number of gas molecules into one side of the box. See Illustration A.

You then open a door in the dividing wall. Over time the gas molecules will spread out to both sides of the box. See Illustration B.

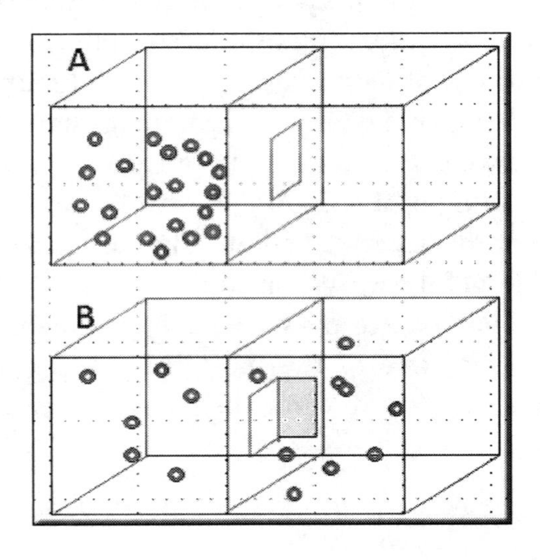

Disregarding the movement of heat, the *other* question is asked: "What is the state of 'order' of the gas molecules in state A and in state B?" The question that has been historically asked is simply, "How easy is it to predict the possible location of the gas molecules in each situation?" or "How 'orderly' are they?"

It is pretty simple to see that the gas in B has more space and can go to a lot more places. It is much more difficult to predict the position of the gas than in A. Therefore the conclusion has been that the gas naturally moves from more predictable order to less predictable order—disorder. It doesn't matter if the box is here on earth, or out in space or on another planet; the conclusion is that the gas molecules will *always* and spontaneously move to maximum disorder. We repeat, the basic concept emerged that the "entropy law" is a law of *increasing disorder.* That moving particles would somehow arrange themselves into more ordered states is "infinitely improbable."

Boltzmann had to work with inadequate information, so he needed to make assumptions about the gas particles. The nature of atoms and molecules was not known at that time. In 460 BCE Democritus thought that a smallest possible bit of matter existed and he called these basic matter particles atoms. Yet more than 2000 years passed before people once again looked at the structure of matter. John Dalton in the 1800s did experiments that showed that matter looked to be made up of small particles. The evidence pointed to something elementary. But years after Boltzmann's work, J. J. Thomson in 1897 created a model of the atom. Boltzmann, however, in his thought experiment imagined and modeled his gas molecules as non-reacting or non-connecting billiard balls colliding in a box.

Boltzmann concluded that each collision would increasingly produce disorder.

The second law, Boltzmann argued, was thus simply the result of the fact that in a world of mechanically colliding particles, disordered states are the most probable. Because there are so many more possible disordered states than ordered ones, a system will almost always be found either in the state of maximum disorder or moving toward it.

At that moment in history, the mid-1850s, there was little interest in the long term fate of the universe. What was exciting was how this discovery affected the Industrial Revolution! The revolution that promised to take the toil and sweat out of human labor was the steam engine. And what mattered was how to make those remarkable steam engines work better. Boltzmann helped solve the big problem of how to manage steam engine efficiency. No one really cared about the disorder or the box; they did care about hot steam that could move pistons in steam engines. This law, in fact, worked and met the needs of the time and there was no incentive to examine its implications further.

A few words in the conclusion of Boltzmann's 480-page book on gas theory set in stone the supreme law for the future of humanity that would be believed for the next 150 years!

For us to be as clear as possible we tracked down a translation of Boltzmann's *Lectures on Gas Theory*, written in 1895. It is comprised of 480 pages of extremely intricate and complex mathematical equations and on page 443, in conclusion, he writes, "one must assume that an enormously complicated mechanical system represents a good picture of the world, and that all or most parts of its surroundings  are initially in a very ordered – therefore very improbable – state. When this is the case, then whenever two or more small parts of it come into interaction with each other, the system formed by these parts is also initially in an ordered state, and when *left to itself it rapidly proceeds to the disordered most probable state.*" (Emphasis ours.)

It is simply wrong.

In the same book Boltzmann added at the end, "Who knows whether they (the concepts in the book) may not broaden the horizon of our circle of ideas, and by stimulating thought, advance the understanding of the facts of experience."

By "stimulating thought" we find that entropy as increasing disorder is not just wrong, and that "advancing understanding of the facts of experience" reveals the *opposite* activity of increasing order.

Heat became the easy way one could detect any change in the arrangement of the gas particles in the box. Heat could be *measured.* Thus with many experiments it became clear that heat would spread out in any space that was enlarged. So the idea came into being that the spread of heat energy would also naturally result in the increase of disorder.

Since the movement in all cases was disordering this idea moved beyond a "theory" or "principle," it became a universal scientific law, taught in every high school and college science class and probably quickly forgotten. Even Boltzmann's first assumption, "one must assume that an enormously complicated mechanical system represents a good picture of the world," has long been discredited by discoveries in relativity and quantum physics. It now is known that the universe is *not* in any way like a giant mechanical clock or any other mechanical system.

This "disordering" was pointed out to one of the authors very clearly when working with a company that made electric motors. George mentioned that we were very uneasy with the second law of thermodynamics. On a break one of the senior scientists walked out on a balcony with us that overlooked the factory floor. He said, "Look down at those machines winding wire around the core of an electric motor. The machines work hard, they get hot, they rust and they wear out. That is entropy at work." Jack, our client, was around sixty years old at that time and he added, "And our bodies too. We just wear out." We'll come back to that factory later and see how entropy thinking permeates everywhere.

Newton's breakthroughs in the late seventeenth century promised an era of science: of answers to the great mysteries of life. The revolutions that followed gave so much hope that science began to replace religion as a foundation for belief in a benign universe. But as science moved ahead new questions emerged. Science was getting no closer to the deeper search for meaning. It could tell us how things worked, but not why.

And Boltzmann pointed to a dismal future.

Where does the second law of thermodynamics leave us? Either with a purposeless universe or a destructive purpose. Is it really any wonder that the twentieth century was marked with a dismal beginning of pessimism and cynicism, and nihilism? The concepts pervading this period were the meaninglessness of life, the lack of any objective value or purpose, and that morality and ethics do not inherently exist and are artificially contrived.

In some respects the philosophy of the twentieth and twenty-first centuries is that deep reality and meaning does not exist. This promoted a general mood of despair around the pointlessness of existence and the lack of any real rules or "norms" or laws. The philosophy of Camus, Kierkegaard, Nietzsche, and Heidegger has been popular; existentialism and postmodernism got the podium.

Even with the advent of the Age of Science, Newton and the flood of discoveries that ensued left many unable to reconcile their religion with an age dominated by science. It was a time that envisioned a deity who could thrust our universe into an ultimate chaos and heat death. A "heat death" would result from the loss of heat as disorder and chaos cooled the cosmos beyond anything one could imagine as cold. People even questioned whether this kind of a deity could be trusted or believed.

All manner of spiritualism and alternative deisms abounded, many flowing in from obscure, esoteric and sometimes reprehensible sources. A tide of frustrated spiritual seekers arose.

This continues today. We can still see and feel the sense of pessimism and negativity today: on the streets, and even on our newscasts where in the face of any good news there is always a "but" to offset anything positive, lest we start to feel good about the direction of things.

The meaning and fate of the universe is simply too important. As Beth said on the flight back from Ireland, "There is just something fundamentally wrong about what Paddy was telling us, with what science accepts about the whole idea of entropy." We decided to revisit this very basic box and gas experiment and to look at it a different way. What could possibly be those other "laws of physics" that Erwin Schrödinger and the others have mentioned?

When Einstein was working on the theory of relativity one of his enormous breakthroughs came from imagining riding on a beam of light. Now that we know more about atoms and molecules, we decided to try his method: to imagine riding on one of the gas atoms in the box.

We did just that, and found that the atoms shook to and fro like a multidimensional bumper car, moving unpredictably from place to place, vibrating and rotating and trying to buck us off. The atoms were pushed and pulled around by what is called Brownian motion[20] and moved in random directions and, of course, they were affected by gravity. We

---

20. The random moving of particles suspended in a fluid a liquid or a gas resulting from their bombardment by the fast-moving atoms or molecules in the gas or liquid.

ran into other atoms and also bumped into the sides of the box.

We were flabbergasted at what we found, to say the least!

CHAPTER FOUR

# THE INVISIBLE BOX

We suddenly realized that in all of the calculations that have been done around these gas atoms enclosed in this box, *the walls of the box have never been included*—even though a box is absolutely required to make the experiment, and the walls of the box contain millions, if not trillions, of atoms and molecules.

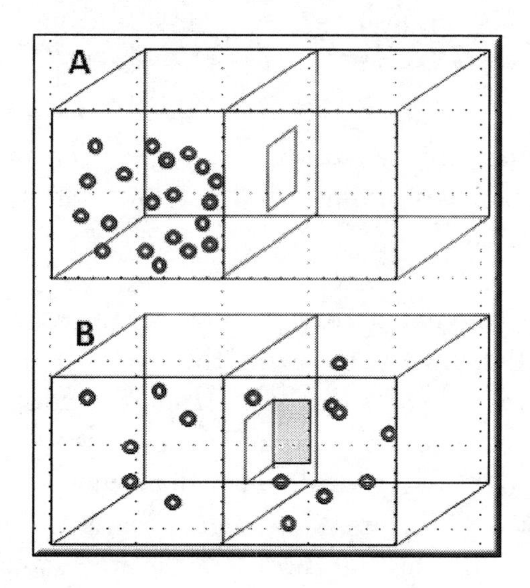


*As seen in the previous diagram (repeated here), the Boltzmann box. The walls of the box are invisible. There is nothing with which to connect or react.*

The simple gas molecules and atoms in box B have a much

greater probability of connecting in new ways and forming new, more complex, connected arrangements.

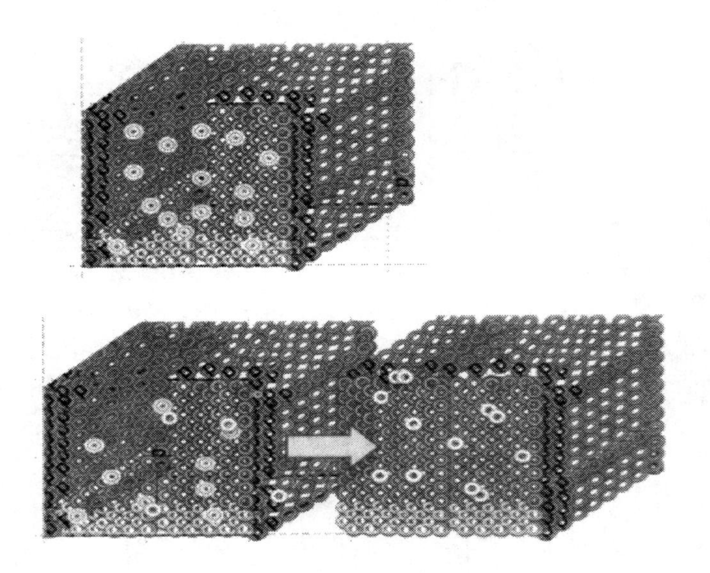

*The grey circles are the box atoms and the white are gas atoms. Now we see the vast number of new atoms and possibilities of new connections.*

What a different picture! We now see the other atoms and molecules that always come with the box(es).

It becomes obvious that to expand or to add another box spreads out the gas but also adds countless new atoms and molecules in the new box, all of which provide *new opportunities for new connections.* Also in the equations leading to the conclusion of increasing "disorder," the atoms and molecules in the walls of the containers have never been included—or the new connections that are then made possible!

All atoms are interactive. The gas is made up of perhaps millions of atoms. The walls of the box contain (solid packing) material which actually has trillions of atoms. In the experiment, Boltzmann held constant the number of gas atoms, but with the new box added, he doubled the number of atoms in the walls and never included them in his equations. The number of possible connections multiplies enormously when all the atoms are included.

Why has the box itself never been included in any of the equations? Boltzmann merely mentions that the box is "perfectly elastic" and non-reactive. No such "non-reactive" material exists in the real world! Of course the experiment cannot be done without the boxes.

The solid walls of the box itself are made from some material that already contains connected atoms and presents innumerable opportunities to bond with the gas. Every atom in the world is also reactive to some extent and can make a variety of different connections.[21] With the door opened to expose more walls, the gas atoms have twice as many new molecules and atoms to explore and with which to bond beyond the gas molecules themselves.

In our study of contemporary papers we found no mention of the container save in one article that mentions a "box-like container" and then suddenly drops the subject. In the translation of Boltzmann's work we also found other assumptions about the box experiment. Boltzmann's assumptions were quite casually dropped into his manuscript at various places. As we noted earlier, at that time nothing was known about atoms so all he had to work with were his assumptions – most of which were wrong.

**Assumption 1**: Gas atoms are "invariably perfectly spherical."

**Reality**: Atoms and their electron "clouds" take on a variety of shapes.

---

21. A large variety of bonds can happen: for example, hydrogen bonds, van der Waals bonds, London, electrostatic, ionic, or covalent.

**Assumption 2**: Gas atoms are "invariably perfectly elastic" (*i.e.*, they do not bond with anything – just bump off).

**Reality**: Gas atoms actually react and bond in a wide variety of ways.

**Assumption 3**: Gas atoms are "not rotating or vibrating."

**Reality**: Gas atoms rotate extremely rapidly and continuously.

**Assumption 4**: Gravity is not considered at all.

**Reality**: Gravity is always present.

What if we take into consideration the real world with real atoms; the reality, if you will?

Of course, opening the door to the larger box would disperse the gas into more "disorder," but if one were to hang around for a while, nature would have the gas atoms and the box atoms bumping into one another and interacting with one another and in many cases connecting with one another—creating connections and new "order."

It is true in the box experiment that at first when the gas moves into the larger box it spreads out into more disorder. But then, over time, the gas begins bonding in a variety of ways with other gas particles and then with the material in the box itself. So, if you will, Boltzmann is half right—in a snapshot.

Frankly, what astonishes us is that a question dealing with the overall direction of the universe and answered by a law embraced for a century and a half has not been reexamined in the light of facts discovered over the past hundred years.

The universally accepted answer has been that the universe is running downhill; that chaos is the end result of the universe; that our future lies in "disorder," in a heat death. If you Google "heat death" you will get millions of results.

This accepted answer has not been challenged. It is taught in every science classroom in high school and college in the nation and probably in the world.

Amazing!

This universally accepted answer is simply wrong. In the real world, the universe is getting more connected, more ordered, more interdependently linked.

Boltzmann concluded that gas atoms never bond, but William Lewis in 1916 and 1918 qualitatively proved how reactions happen when particles of gas hit each other.[22] A certain percentage of the collisions always cause significant chemical bonding.

It's very hard to find actual experiments done on the second law; however we were able to find a real—not a thought—experiment. It was reported in a 2008 *Scientific American* article, "The Long Arm of Thermodynamics."[23] J. Miguel Rubi and his colleagues at the Norwegian University of Science and Technology reported on their research on the second law. They concluded, "The second law does not mandate a steady degeneration. It quite happily co-exists with the spontaneous development of order and complexity."

According to the accepted second law, the universe must have started out in its most orderly state and has been getting messier ever since. This just isn't happening. Disorder is constantly being reduced.

If this kind of spontaneous relationship building did not happen in nature, the evolution of the universe from a vast cloud of gas into a collection of diverse types of matter would never have happened. The universe would still be wandering around as a gas, becoming more and more dissipated and disordered.

Remember our experience on the factory balcony where we were shown that the machines were wearing out. We

---

22. M. Nic, M. J. Jirat, and B. Kosata, eds., "Collision theory," *IUPAC Compendium of Chemical Terminology* (Online edition).

23. J. Miguel Rubi, "The Long Arm of Thermodynamics," *Scientific American* (November, 2008).

visited that same plant five years later. The factory floor was very different. The machines were different. Half the factory floor was empty. We asked about this and the explanation was that machine operations had been consolidated and the step preparing the wire was incorporated into the wire wrapping machine. “It’s faster, more efficient, uses less energy and gives us a real advantage over our competition.”

In the box and the factory examples of entropy, we and others have made what we now call the “content, context mistake.” The entropy law came into being without considering the whole environment, only the imaginary box. Just adding in the molecules in the box totally changes the mathematical conclusion. When you consider what happens over time even more changes occur. The machine example had not taken into consideration the factory and the natural development over time of better, more efficient and connected machines – the environment.

To really draw conclusions about anything we have to consider not just the thing, but its environment (the system) and how it changes over time.

By the way, look in your garage. Over the years some other car has replaced the one that had declined in performance and in some cases was falling apart. You had to replace the old car with a new one that connects with you and your environment in better ways, from using less energy to being more comfortable and even connecting with a satellite 12,000 miles above the earth to tell you where you are and how to get where you’re going. You’re getting more connected. You are connected with the world with your computer and your television.

That’s what Nobelist Albert Szent-Györgyi called “syntropy”—the opposite of entropy: becoming more ordered, more interdependently connected with the environment. It doesn’t go backwards returning to some old order; it goes forward making new connections and most often at a more complex level of organization.

We are often fascinated by popular science television shows that illustrate entropy. One illustrative device involves breaking an egg and then running the film backwards to show the egg reassembling itself back into the original egg, pointing out that this could never happen. They do not show what happens to the egg going forward. Let's look at the usual case.

Most often the egg is broken for a reason: to cook it and eat it. So it is taken into the digestive system, down to the gut, where it further disassembles, creating more disorder. It is absorbed into the body, and miracle upon miracle, its molecules join in the dance to become part of living cells in the most complex organism we know of—the human body. Some parts of it have gone on to become effluent and play in other connecting dances. They never go backwards to the old egg. It's a good idea to look at the whole activity over time, not just one little part of it in a snapshot.

Everything around us is in the connecting business. You open a soft drink can and hear the little fizz. The gas had been dissolved in the liquid and has been released. It's certainly more localizable and predictable when it was in the can than when it goes out into the surroundings. But what is its future? Now the gas enjoys innumerable opportunities to bond with other atoms and bring about new connections and order.

It happens everywhere. On the human side, imagine moving into a new home. The movers put the furniture all over the place. No matter how hard you try you will not be happy with how things are left; you will rearrange the furniture to suit you. You want everything to be in a good relationship with the other things around it.

Flash forward and you've bought a new sofa and table. There you go, rearranging your new furniture in order for the items to "fit together." Well, you say, "That's just natural."

And that's just what we find in nature. Nature connects things in new interdependent organizations and as things change, *it will reorder.* We call this intermediate state of reordering "chaos." At least that's what some in the family say when we start doing our rearranging.

Anyone, for example, looking at America in the mid-1770s would have predicted that the various states, the agricultural slave states in the South and the industrializing states in the North with all their differences would never agree to join together. Looking at those poor delegates sweltering through the summer in a six-week debate in Philadelphia would just make them shake their heads in despair. But by collaborating and creating, they formed the American union. They were open to the creative connecting purpose of nature.

How many of us, or someone we know, have experienced a major disaster in their life: a flood, loss of a loved one, a lost job, an accident?

Many research studies show that in the vast majority of cases, things get better. We get more connected: with ourselves, our families, our friends and our lives. Studies find that a natural disaster can actually increase the long-term level of economic output. This is consistent with the idea of "creative destruction" as an important determinant of economic growth. A natural disaster, by destroying the social and economic institutions that were impediments to economic growth, often enhances economic activity.

Ecology teaches us that forest fires can be a form of creative destruction or restructuring. Fires create openings in the forest canopy that allow sunlight to reach smaller plants. This stimulates ecosystem diversity. Burning also enriches the soil by depositing calcium, potassium, phosphorus, and other minerals.

Arthur M. Diamond Jr. of the University of Nebraska at Omaha noted that in economics, creative destruction is the process by which entrepreneurs introduce innovations that force established businesses to adapt or die. The phrase "creative destruction" was coined by the eminent economist Joseph Schumpeter (1883-1950), who believed that long-term economic growth is stimulated and sustained by radical innovators, even as they destroy the value of large, dominant firms.[24] We would prefer to call it "creative restructuring." Creative restructuring opens the economy to increasing the rate of economic growth, as well as increasing the length and quality of human life.

A recent and obvious example of creative destruction is the shift from first-class mail to email. In addition, the internet has created countless examples of creative destruction. Ebooks have just about eliminated book stores; video rentals via the internet have had the same effect on video rental stores. The amazing popularity of Facebook, Twitter, blogs and other social media indicates the rise of new connecting phenomena. In each case the underlying appeal is to stay connected and build more connections.

Let's get back to the second law of thermodynamics. How could this kind of connecting process have been missed in the original formulation of the second law of thermodynamics? At the time entropy was defined, atoms and molecules themselves were not well known or understood, much less the ways that atoms bonded to form new molecular order. Because of this, we can more easily comprehend the misunderstanding.

The leading scientists of that time *did not understand that atoms actually existed.* It's fascinating to note that the Greek philosopher Epicurus not only theorized atoms and their abundance but that everything that happens results from atoms running into each other and becoming intertwined

---

24. Joseph A. Schumpeter, *Capitalism, Socialism and Democracy* (New York: Harper & Brothers, 1950).

with one another. Epicurus's atoms occasionally exhibit a "swerve" away from straight lines, thus avoiding the rigid determinism implicit in other ideas[25]—all very close to today's thinking. It's no wonder that Einstein kept referring to "the ancients'" wisdom.

The verification of atoms came years after Boltzmann's death, but the second law was not revisited. Also, as we mentioned earlier, it seems as if this law of entropy was one of those magical illusions. All of the attention was focused on the content—the gas molecules and their temperature. The movement of heat was and has been like a great engine pulling the second law forward with the idea of entropy riding along as a caboose, an unexamined assumption.

As Albert Einstein put it, "The theory decides what we can observe." Even with the modern understanding of atomic bonding, all the entropy equations have been built upon and modified in such a way that any student looking at entropy today would just see a universal law mathematically expressed. Gases and heat transfer are represented as symbolic equations—focused totally on the energy, the heat in the system's parts ("content")—without any reference to the "context," the connections among the parts or the environment that make up the system. Mathematics, as usual, focused directly on the *quantity* of things rather than the *quality* of the occurrences.

Practicing science today means building on the discoveries of the past, not going back to how the original conclusions were reached. The box, the environment, is lost in history. The original interpretation of entropy energy and heat behavior does contribute to scientific progress. After all, if all you are looking at is the movement of energy, the basic

25. Lucretius, *The Nature of Things*. Translation by Frank O. Copley,(New York: W.W. Norton, 1977).

second law, just known now as some equations, is perfectly satisfactory in describing the movement of gases.

We just miss the point of other processes going on and nature's deeper function of Creative Connecting.

We can't explain how it happened, but it seems that somewhere along the way of thinking about how heat always disperses, Boltzmann and Clausius and many others concluded that the movement of heat could also be carried over and applied to matter! This, after all, was a thought experiment with no opportunity to follow the later behavior in the experiment. And, over many years, no follow-up experiments occurred. When we add it all up, we find that all of nature clearly appears not to be coming apart as the original second law claims, but on the contrary, is pulling together, connecting elements in the environment – pulled by the purpose of creatively connecting and unifying the cosmos.

Let us state again, what we see is that the universe has a grand purpose of connecting, of unification, of bringing everything in our world together. Starting with the basic, simple overall connecting force of gravity, nature proceeds to pull the universe toward ever more and deeper, more complex, interdependent connections.

The noted physicist, Henry Stapp, described the facts of physical reality to the Atomic Energy Commission as "a web of relationships between elements whose meanings arise wholly from their relationships to the whole." Everything—including us—exists as sets of relationships. How anything relates to those things around it determines what it is!

Erwin Schrödinger, one of the fathers of quantum physics, in his 1944 book *What is Life?*, saw life struggling against the power of entropy and concluded: "It emerges that living matter is likely to involve '*other laws of physics*' hitherto unknown, which, however, once they have been revealed, will form just as integral a part of this science as the former" (emphasis ours).[26] The general concept of positive entropy and the phrase "negative entropy" were introduced by Schrödinger and later Léon Brillouin shortened the phrase to *negentropy*. In 1974, Albert Szent-Györgyi endorsed replacing the term *negentropy* with *syntropy*, a concept and word originated by the Italian mathematician Luigi Fantappiè in the 1940s.

This connecting force touches our lives in ever so many ways, from the ways we form relationships, to putting objects together in new ways to accomplish new functions, or mixing up that new batch of cookies with that secret ingredient to surprise the kids. We are "creative connectors." Just look at the many "mundane" things we do in the course of a day and see how many of them contribute to "connecting." How about doing the wash? Fixing a meal? Taking the kids to school or sports? Meeting a customer? Sending an email? Why do we pick a particular article of clothing if not to present ourselves well and connect with others in a good way? (Of course it would be nice if it fit, too.)

What we observe in nature is that at each step in complexity, nature creates mutually supporting processes where connections increase. In addition, following one of the basic foundations of the creative process—creating many possibilities—the universe creates many, many alternatives, from planets to atoms to molecules to cells to plants and animals and on to humans, tribes, societies and nations. As we look for the source of creativity we find that nature incorporates creative divergence, the generation of a wide variety of alternatives, as the bedrock in its most fundamental activities.

So many different expressions of life have been tried that it is estimated that over 98 percent of plants and animals have

---

26. Erwin Schrödinger, *What Is Life? with Mind and Matter and Autobiographical Sketches* (Cambridge University Press, 1992) [First published in 1944]..

gone extinct; nature chooses to move ahead with those that are "most fit." This term "most fit" has been used without clear definition since the term was first introduced. The term requires the utmost clarity.

One of the ways the fit are selected is by means of a pruning process. In the pruning process in nature, those that survive clearly are those that connect best with their environment, those that connect and share energy in the most efficient way. The "most fit" absorb energy and materials and interact with their surroundings in mutually beneficial ways. That is, the "survivors" tend to be *cooperators* (better connectors) with those around them.

All the recent studies have moved away from the idea of "strongest" to those that practice "mutualistic" symbiosis. Mutualism makes up a form of symbiotic relationship where participants on both sides benefit. For example, on the roots of almost all trees there is a fungus that provides the tree with minerals from the soil, while the tree provides carbon from the atmosphere. Neither could survive without the other. These kinds of mutualistic bonds make up the basis of all healthy ecosystems.

Most bonding or connecting in physical systems is clearly understood and makes up the basis of physics and chemistry. Some remain enigmatic—like water. The largest supercomputers still cannot model the amazing cooperative behavior of water, the most vital element of life.

All of the most important natural connections defy any "logical" explanation or mathematical equation. And, again, each has those special characteristics of creativity:

- Unpredictability
- Radical novelty
- Connection between dissimilar things
- The creation of a new, integrated whole

- A new whole that is greater than the sum of its parts
- A new whole that is is different than the parts
- Often, greater efficiency in sharing the physical energy available
- Increased connections with the environment

As we discovered, there is no math nor are there equations to account for creative or emergent phenomena. Nature abounds with emergent phenomena, including the appearance of life itself or subatomic particles, forces and atoms created from the nucleosynthesis of supernovae.

In the human domain, emergent phenomena are called "creative." The telephone, movable type, electricity, the light bulb, the steam engine, ebooks, radio, television, airplanes, smart phones, antibiotics, paintings, music, films, are creative expressions fashioned without precedent (excluding sequels). Each of these examples couldn't have been predicted based on the past.

Science based on mathematics cannot account for these radically unique evolutionary events. We are dealing with great changes in quality, in many properties, not merely quantities. We cannot predict or explain a butterfly, a walrus, an automobile, or even the cyber world. And the new expressions are unique, not patterned on the past—or on logic. The biosphere and humans keep inventing brand new stuff that behaves in distinctive ways. We can't say ahead of time what they will be. Yet, we are taught that logic and reason can answer all questions. Our experience with "experts" in many fields shows that they often tend to point out from past experience or knowledge why some idea won't work. It's just not reasonable "Sending pictures through the air … how ridiculous!"

This presents a formidable, seemingly insurmountable challenge to physics and the deep logic mythos that surrounds both science and mathematics. Physics cannot explain life. For example, many attempts to explain nature, particularly in the most modern science – quantum physics – end up logically with infinity. But *creativity always leads to infinity.* Computers attempting to predict the future are painted into a corner as they depend on past facts. They cannot create the unpredictable or novel.

Picasso or Longfellow or Edison or Ford were not worried that their thinking could wander infinitely, but scientists have resorted to tools that can change natural infinites to real numbers, requiring a mathematical technique called "renormalization." Noted scientists have quite unsympathetically questioned the validity of this treatment or renormalization of infinities.

The Nobel Prize winning physicist Paul Dirac said:

> Most physicists are very satisfied with the situation. They say: "Quantum electrodynamics is a good theory and we do not have to worry about it anymore." I must say that I am very dissatisfied with the situation, because this so-called "good theory" does involve neglecting infinities which appear in its equations, neglecting them in an arbitrary way. This is just not sensible mathematics. Sensible mathematics involves neglecting a quantity when it is small - not neglecting it just because it is infinitely great and you do not want it![27]

Yet an additional important critic was another Nobel winner, Richard Feynman. He reflected:

> The shell game that we play ... is technically called "renormalization." But no matter how clever the

27. Helge Kragh, *Dirac: A scientific Biography* (Cambridge University Press, 1990), p. 184.

> word, it is still what I would call a dippy process! Having to resort to such hocus-pocus has prevented us from proving that the theory of quantum electrodynamics is mathematically self-consistent. It's surprising that the theory still hasn't been proved self-consistent one way or the other by now; I suspect that renormalization is not mathematically legitimate.[28]

Giordano Bruno, an Italian Dominican friar, philosopher, mathematician, astronomer and cosmologist, suggested among other things in 1600 that there were such things as other intelligent life in the cosmos and infinities. Bruno's claim that the universe was infinite and intelligent disagreed with the church's dogma. He was burned at the stake because he would not recant.

At some point we must realize that the creative process, by its very nature, contains both infinities and the "illogical," going beyond today's mathematics and rationality. We need to rethink what we believe about nature, its purpose and process and give "creative thinking" and imagination the kind of respect, inquiry and attention that we have given to, and been limited by, mathematics, reason and logic.

Truly, we have not taken full advantage of the infinite possibilities inherent in our abilities—gifts endowed to everyone and everything in the universe.

---

28. Richard P. Feynman, *QED: The Strange Theory of Light and Matter* (New York: Penguin, 1990), p. 128.

CHAPTER FIVE

# NEW POSSIBILITIES

---

We pause here for an important announcement.

Psychologists call it "selective attention": paying attention to one thing and often being blind to something else right beside it. As Albert Einstein put it more simply, "It is the theory that decides what we can observe."

In the midst of writing this book, we were stopped in our tracks!

We had been concentrating so exclusively on the idea of *connecting* and *order* that we had casually and happily skipped over any phenomena of *disorder.*

The disorder part of nature's activity just didn't fit in the theory that we had formulated about the building up of order and connections. How could disorder fit into building a higher order of connections?

We had trapped ourselves in our own theory. We were putting together a theory and by God we were going to stick to it! We had made the very mistake that Einstein had warned against: we had seen only what our theory had allowed us to see.

We went into a deep funk.

Why were we affected so profoundly? While writing these pages we were engaged in two conversations: one with you on these pages, and the other between ourselves about the meaning of what we were discovering and what it meant to us personally.

With our new understanding of entropy, a new meaning of

the idea of "connecting" had emerged. It had touched the edge of what we felt is the life of our spiritual world. We had always been spiritual seekers, deeply unsatisfied seekers. As we pondered this notion of connecting, connecting with everything – with knowledge, with objects, with community, with people, with family, with nature, with a sense of connectivity with the cosmos itself – we discovered that it was the *connectivity itself* that brought us that sense of peace and joy. Our spiritual life was revealed! Epicurus, who was born in 341 BCE and noted as the father of the scientific method, believed that communing with friends was the greatest of all happiness.

That other conversation was very exciting. It said to us that we had discovered a fundamental natural law that revealed a universe designed so that we could become unified. Our destiny is to become ordered, not to become torn apart in cold chaos.

Suddenly we were facing a bitter pill. We had to admit the existence of disorder and randomness that the quantum physicists had proven. And that it somehow must fit into our theory.

Once again we had to go back to source, because the evidence of nature's creativity was all around us. The wonders that surround us are not a mirage. There must be a way!

Our depression lasted ... for a while.

And then lightning struck!

It turns out that disorder is absolutely essential to achieving nature's overall design. Our bitter pill, like many in life, turned out to be extraordinarily important.

At this point we ask you to join us on our journey to escape the boundaries of some past assumptions that have kept us from seeing new possibilities. We had taken a path up to the top of a hill from the flatland where we had started, and suddenly we were presented with an array of not only the connecting and ordering forces but their *interaction* with those other forces of disordering and disconnecting.

We will have two main partners on this journey. One is your imagination and intuition. The other is Albert Einstein, who noted that "Imagination is more important than knowledge" and "Intuition is the only thing that is real."

When we talk about Einstein don't get glassy-eyed; we will not be involving you in complex theories or equations, but with the profound *meaning* behind Dr. Einstein's thinking as he formulated his theory of special relativity. It can lead us to a resolution of the order-disorder question and the bigger question of "meaning" – how anything has meaning to us, from the relationships we have with others to how we feel about the meaning of our lives and the link between science and spirituality.

Why would special relativity or the second law explain why we feel so good when we make a new friend, or so bad when we lose one? Why do we feel a little jolt of pleasure when we solve a puzzle or even just "connect the dots?" We intuitively feel how the universe works. In this feeling of connecting we are feeling the source of creativity.

First let us set up a little background. As far back as anyone can look there has always been an interplay in human thinking between the greatest forces of nature: God and the devil, good and evil, yin and yang, *eros* and *thanatos*, life and death, heaven and hell, light and dark, Brahma and Shiva, male and female. In the physical sciences we could go on with the entire world of positive and negatives often first thought of as opposites and realized later to be complements. As the ancient Chinese once said, "This arises, that becomes." With light, you have darkness; one demands the other. The positive charge of an atomic proton combines with the negative charge of the electron. Atoms with a surfeit of electrons merge with atoms with a deficit; cells in our bodies die and are replaced with newborn cells to keep us healthy. Death produces life.

Einstein made the discovery of the century when he found

that enormous energy was stored in matter in the theory of special relativity, $E=mc^2$, producing the promise of atomic energy. This may seem a very long way from what this book is about, but please stay with us. Einstein discovered much more than any of us imagined. Nuclear energy may—in the end—be the least of it.

When we first started this book we were talking about it with a good friend, Othón Canales, and he asked, "What are you going to do about $E=mc^2$?" We had a good laugh about that. Now, his comical comment has come back to roost. Remember when we talked about Boltzmann's entropy concept being hitched on to his work about steam power? Well, a very similar thing happened with Albert Einstein and his theory of special relativity.

The overwhelming breakthrough that changed the world forever was $E=mc^2$. It won a war. It held out the promise of infinite energy. It made a giant leap into a new understanding of nature. For a while it overshadowed the brilliant advances of the physics of quantum mechanics that was also going about the business of changing our world with the digital revolution. But Einstein hated quantum mechanics!

The reason Einstein took such profound issue with quantum mechanics was that in the series of equations that produced $E=mc^2$ lay an equation that required Einstein to explain the *why* behind it all, that is, what caused it to happen at all. What was the moving force that caused this energy to be buried in matter? The mathematical equation that led up to $E=mc^2$ was absolutely required and it is a peculiar animal, this equation. *It produces two correct answers*—as you will see.[29] One answer says that causes rely on determinism on the clas-

29. $E^2=m^2c^4+p^2c^2$, where E denotes energy, m denotes mass, c denotes the constant of the speed of light and p denotes momentum. Einstein proceeded to simplify the equation, resulting in the famous $E=mc^2$, which has only a deterministic solution.

sical concept of cause and effect, that every effect is preceded by a past and discernible cause. The other answer is the opposite: the cause is in the future and is not predictable.

In classical mechanics, a particle has, at every moment, an exact position and an exact momentum. These values change deterministically as the particle moves according to Newton's laws. You can know exactly where a particle is and how fast it is moving. In quantum mechanics, particles *do not have* exactly determined properties. When you look at where the particle is, its position, you automatically lose the ability to know how fast the particle is moving. And when you measure how fast it is moving, its momentum, you can't find out exactly where it is. In the ordinary world, we watch runners and we can see both where the runners are and how fast they are running. And you could look to the past to figure out where they were and how fast they were going. Not so in our small world of quantum physics (of which everything is made): you could see the runners but you could not tell how fast they were running, and if you measured how fast they were running you could not tell where they were! Where does this strange uncertainty come from? If you know one thing, you can only guess about the other. Before quantum theory, physicists could believe in determinism, the idea of a world unfolding with precise mathematical certainty. Since then, however, the weird probabilistic behavior of the quantum world has rudely intruded, and the mainstream view is that this uncertainty is a fundamental feature of everything, all the way from the littlest alpha particles to the biggest of these little things.

So when they are measured, the result has to be randomly drawn from a probability distribution. This gives you what is called the *non-determinism* and seeming randomness of quantum physics – the seeming arbitrariness of the universe. The universe will not let you know both, position and momentum, at the same time. As physicists and others say, "It makes no sense."

The theory of quantum mechanics has puzzled everybody but has won over the physics community with its vast body of evidence and practical uses. Quantum physics proves beyond any doubt that the future is just "a roll of the dice"; it is essentially "probabilistic," pretty much random—*not* determined by the past.

For Einstein and many others, quantum physics was a bigger bombshell than the atomic one. It had destroyed any kind of an intelligent or even rational God! It had proven a universe without any meaning or purpose or any reason to be; just some kind of cosmic accident. No wonder Einstein was upset. But, in the end, Einstein buried a deterministic equation in his relativity theory. In his view, the past still produced the future! Cause would still produce effects.

So Einstein believed in a deterministic universe and because of the proofs of quantum mechanics the world of science now believed in the other view—one of randomness. For us mortals, what's left for us to think about here, other than the basic question of how the universe works and what on earth we can believe in? Can science ever uncover evidence of a God? Is there a God? If there is, what kind? Why are there human beings? Why should anything happen? And so on. Pretty good questions.

And there are answers.

From unexpected places.

We can find answers in Boltzmann's work, special relativity and quantum mechanics—each, as we will see, make up a piece of the great puzzle, but only when they are joined together. The great jigsaw puzzle is how to make meaning from such seeming opposites as the randomness of quantum mechanics and the order of determinism.

On one side of the facts is Einstein, who says that there *is* a cause behind everything. And proves it with $E=mc^2$; because buried in the equations underlying $E=mc^2$, it is possible to find determinism right there, hidden in the math. And nuclear power works, doesn't it? On the other side is quantum mechanics, with thousands of experiments that work exquisitely well, and that absolutely verify the theory of uncertainty and randomness and no past cause. It turns out *both are right.*

We are seeing another case of *complementarity.*

There is little wonder that the physics community does not discuss these subjects. When you start talking about randomness and uncertainty, most scientists head for the exits. This is pretty much left to philosophers, who by and large have little knowledge of the deep esoteric notations of mathematics, and to the physicists who overall agree with the world famous physicist, Richard Feynman, when he merely says, "Quantum physics simply cannot be understood." As one senior researcher said to his young colleague who kept pestering him with the question of why quantum mechanics works: "Shut up, and calculate."

We are going to look at this whole paradox in a new way: an understandable way. The seeming craziness of quantum mechanics will make good sense. Albert Einstein urged us on when he said, "To raise new questions, new possibilities, to regard old problems from a new angle, requires creative imagination and marks a real advance in science." This is what we are doing.

When Einstein was creating the final equations for special relativity he ended up with what are called "quadratic equations." Oh boy, we're sure this gets you readers very excited! Well, this will take a few minutes of very basic "mathematical" thinking. In this case, it turned out to be very important—but actually it turned out to be pretty simple math, believe it or not, so don't get worried. We promise you this little bit of figuring will be worth it.

It will actually get down to two times two. OK?

If you care about getting to the bottom of how you actually *already know* what is the profound driving force of the universe and how important it is to feel our unity with the universe, then you will end up appreciating this little detour into a fundamental question in science and the one we are looking for here – "What is that force in the universe that drives creativity in nature and humans? And how can I use it to make my life better?"

In the end, you will find that the longing we have all shared from our hearts and minds to be able to join in a profound understanding of a humane cosmos will be worth it. The seeming complexities will dissolve and we will appreciate nature all the more.

Remember what Einstein said, "I have deep faith that the principle of the universe will be beautiful and simple."

This is exactly what we found!

Glassy eyes time again. Stick with it for a little while. And so we move ahead ... We need to consider this thing called "square roots." They are at the bottom of Einstein's quadratic equations. Without understanding the nature of the mathematics of square roots (which is pretty simple) we can never understand the creative force driving the universe.

As you may remember from school, the square root of anything is the number which multiplied by itself gives you the number you are starting with. Thus the square root of 4 is 2, that is, plus 2 multiplied by plus 2 gives you plus 4. Now, the little twist in this square root thing is that a *minus* 2 multiplied by *minus* 2 also gives you plus 4.

To clarify multiplying two minuses – if I just say "Eat!" I am encouraging you to eat (plus), but if I say "Do not eat!" I am saying the opposite (minus).

Now if I say "Do NOT, NOT eat" (two negatives), I am back to a really roundabout way of saying "Eat." As a result, two negatives make a positive. "Do not, not eat."

Therefore no matter how you look at it there are *two* answers to the square root of four: +2 and –2. And both are correct!

This presents a quandary. If you have an equation with an answer that has a square root, like those "quadratic" equations of Einstein's, you end up with square roots and two right answers like we got up above. What do you do in this situation? When you have a math problem you should just get one answer. If you have two, which answer do you pick? And why would that be important anyway?

This was critical for Einstein as his answer to special relativity came as a "quadratic" equation with square roots with *two answers.* The choice of which answer to use was extremely important. If he selected +2 (the positive answer), that had a deep meaning. It meant in mathematical terms that you are describing something that happened in the past; it was deterministic, meaning that all events covered in the equation are determined by preceding events and can be predicted by a past cause. If he selected -2 (the negative answer), it meant that all events are determined instead by some future cause and are unpredictable (future pull).[30] This other kind of causality comes from the unknowable future: something in the future pulls you toward it. It is "teleological" (moving toward some final end that is causing the events). There is some "purpose" behind everything. Einstein came down on the side of determinism, the classic, traditional answer that cause always precedes effect.

---

30. A number of important physicists have seriously investigated the negative solution, including Paul Dirac, J.W. Gibbs, Ulisse Di Corpo, Antonella Vannini, and Luigi Fantappiè.

Perhaps, more than anything, Einstein disliked the negative answer because of its indeterminacy. Its seeming unpredictability would support quantum physics. He liked the rational order of determinacy. He literally hated quantum mechanics. He said, "The more success the quantum theory has, the sillier it looks."[31]

Also, a negative answer would indicate future causality, motivated by some future purpose, and this was perhaps considered the domain of religion. Science was not ready to tread on the toes of religion. We really cannot read Einstein's mind, but in the end he *modified his equation* so that only the positive answer was possible with the famous $E=mc^2$ formula. And it verified the traditional view that everything was moved by and determined by the cause and order of the past.

Two of the founders of quantum physics were Nobel Prize winners Wolfgang Pauli and Max Born. Over the years, they conducted an extensive correspondence. Pauli, writing to Born in Edinburgh from Princeton, noted that in his conversations with Einstein, Einstein really *was not a pure determinist* but a realist, with the conviction that nature is governed by profound levels of intelligible connection that cannot be expressed in the crude terms of classical causality and traditional mathematics.

He (Einstein) was convinced that the deeper forms of intelligibility being brought to light in relativity and quantum theory cannot be understood in terms of the classical notions of causality. They required what Einstein called *Übercausalität* (supercausality). And this called for "an entirely new kind of mathematical thinking." That was a kind of mathematics he did not even know, but which someone must find.[32]

---

31. Albert Einstein, letter to Heinrich Zangger, May 20, 1912. [http://einstein.biz/quotes.php].

32. Antonella Vannini, "Dual energy solution and supercausality," *Syntropy*, 2006 pp. 172-183 [http://www.sintropia.it/english/2006-eng-3-03.pdf].

Perhaps our concept of integrating the double solution to include the negative solution (answer) was not completely absurd. A more recent celebrated Nobel scientist, Ilya Prigogine, concluded that the negative answer should be included. In 1979 Ilya Prigogine showed that the rejection of the negative solution (purpose) of energy in the special theory of relativity has blocked the understanding of the mechanisms that are behind the creative qualities of living systems, which divides culture in two: on one side mechanical science, on the other side life and finalities (purpose). In this way an imbalance between mechanistic science (causes located in the past) and spirituality (causes/finalities located in the future) was established. Prigogine names this imbalance the "*old alliance*."

According to Prigogine, the widening of science to include the negative energy solution will pave the way to a new alliance between science and religion, leading towards a new culture in which science and religion will unite in a new vision of reality. Prigogine names this vision the "*new alliance*," what he calls "a world of possibilities."[33]

Let's go back to the dualities—back to the opposites that become complements. Suppose the perplexing equations that give two correct answers are actually right? Perhaps this is a new kind of mathematics?

James Jeans once said, "God is a mathematician." The Nobel Laureate Paul Dirac noted, "One could perhaps describe the situation by saying that God is a mathematician of a very high order, and He used very advanced mathematics in constructing the universe." Perhaps there is an advanced mathematics that allows both answers to be true? Perhaps this is the mathematics Einstein was talking about – the mathematics of creativity!

---

33. Ilya Prigogine, *From Being to Becoming: Time and Complexity in the Physical Sciences* (W. H. Freeman & Company, 1980).

Einstein said:

> The development during the present century is characterized by two theoretical systems essentially independent of each other: the theory of relativity and the quantum theory. The two systems do not directly contradict each other; but they seem little adapted to fusion into one unified theory. For the time being we have to admit that we do not possess any general theoretical basis for physics which can be regarded as its logical foundation. If, then, it is true that the axiomatic basis of theoretical physics cannot be extracted from experience but must be freely invented, can we ever hope to find the right way? I answer without hesitation that there is, in my opinion, a right way, and that we are capable of finding it. I hold it true that pure thought can grasp reality as the ancients dreamed.[34]

Yet the logical, rational framework of mathematics as we know it cannot now handle creativity or the same phenomenon, "emergence," in physics. It cannot provide an equation for the formation of an atom or a cell or a flower or the emergence of anything different or novel. Yet the universe is constructed of a series of unpredictable "emergences": atoms, molecules, water, cells, life in any form, and consciousness or any human interaction.

The only unifying possibility lies outside of traditional mathematics and in the realm of a creative process that includes both randomness and order.

---

34. Albert Einstein, Address before the Eighth American Scientific Congress, Washington, D.C., May 15, 1940.

But what if Einstein's intuition and Prigogine's concept of "the new alliance" are correct? What if the strange math of two solutions is actually right? What if both answers are correct?

We are suggesting that in special relativity both answers are correct: that the answers are creative complements, not opposites.

The complements are the order of determinacy and the disorder of indeterminacy.

As Prigogine put it, "It lies somewhere between two alienating images of a deterministic world and an arbitrary world of pure chance ... that has slipped through the meshes of the scientific net." He added, "But perhaps there is a more subtle form of reality that involves both laws and games."[35]

Niels Bohr established the principle of *Complementarity*, that is, that a complete knowledge of atomic phenomena requires a description of both wave and particle properties. The brilliant and award-winning physicist, John Wheeler, opined that "Bohr's principle of Complementarity is the most revolutionary scientific concept of this century and the heart of his fifty-year search for the full significance of the quantum idea."[36] We are just beginning to appreciate what Wheeler was saying.

Even Democritus (460-370 BCE) had a great intuition: "Everything existing in the Universe is the fruit of chance and necessity."

We found that researchers have already found practical ways of using this complementary principle of disorder and order. In information sciences, for example, cryptographers

35. Ilya Prigogine, *From Being to Becoming: Time and Complexity in the Physical Sciences* (W. H. Freeman & Company, 1980).

36. Martin Gardner, *Science: Good, Bad, and Bogus* (Amherst, New York: Prometheus Press, 1981).

use random background noise to hide messages. Only those with a key to the noise can get the message. In biotechnology, they have discovered that biologists can find new ways to diagnose and treat diseases like epilepsy and cancer by creating a vast and astonishing array of trillions of random snippets of RNA and DNA and then testing them against the activities of cells. These creative connectors are called "aptamers." Randomness produces new order.

Biologist Joseph Ayers of Northeastern University has experimented with a new way of controlling flying robots. The usual way of handling variations and different situations is to have computer algorithms (*i.e.,* if this happens do this), and that's a difficult problem with so many possible variations in flight. His team came up with the idea of using nature's way—using randomness and chaos to explore many solutions in flight. The helicopter successfully followed the right direction.[37] The same approach is being used to create self-repairing computers at University College London by Peter Bentley and Christos Sakellariou. Rather than use the usual logical program counter to react when, for example, the temperature is too hot, the answer is chosen by a program modeling nature's randomness. As described in *New Scientist*, they reported at the evolvable systems conference in Singapore that it works much faster than expected.[38]

Disorder is producing new order.

Famous Nobel Laureate Wolfgang Pauli, one of the pioneers of quantum physics, says it well:

> There will always be two attitudes dwelling in the soul of man, and the one will carry the other already

37. *New Scientist*, 10 August 2013 p. 9.
38. *New Scientist*, February 2013, p. 21

> within it, as the seed of its opposite. Hence arises a sort of dialectical process of which we know not whither it leads us. I believe as Westerners we must entrust ourselves to this process, and acknowledge the two opposites to be complementary. In allowing the tension of the opposites to persist, we must also recognize that in every endeavor to know or solve we depend upon factors which are outside our control, and which religious language has always entitled "grace."[39]

The former Dominican priest Matthew Fox intuited this creative complementarity long before we did. He said:

> Creativity has an answer. We are told by those who have studied the processes of nature that creativity happens at the border between chaos and order. Chaos is a prelude to creativity. We need to learn, as every artist needs to learn, to live with chaos and indeed to dance with it as we listen to it and attempt some ordering. Artists wrestle with chaos, take it apart, deconstruct and reconstruct from it. Accept the challenge to convert chaos into some kind of order, respecting the timing of it all, not pushing beyond what is possible—combining holy patience with holy impatience—that is the role of the artist. It is each of our roles as we launch the twenty-first century because we are all called to be artists in our own way. We were all artists as children. We need to study the chaos around us in order to turn it into

39. Quoted in Werner Heisenberg, *Across The Frontiers* (New York: Harper & Row, 1974) p.34.

something beautiful. Something sustainable. Something that remains.[40]

All this describes two forces driving the universe: one is a disordering force as Boltzmann found and as we saw in the initial phase of the box experiment as the gas spread out, and the other is the ordering force that we found as the gas connected and reordered itself over time in the expanded box.

Einstein's known past cause blends with the unknown, unpredictable possibilities of the future. The two laws fit together and support each other. Thanks to quantum physics for revealing the disorder, and thus the endless possibilities in the universe for creativity.

Carl Zimmer, a *New York Times* columnist and author of a *Scientific American* lead article, "The Surprising Origins of Life's Complexity," pointed out how a variety of scientists working with diverse organisms found disorder leading to higher order. He notes that Darwinian natural selection has held sway for a long time, "But recently some scientists and philosophers have suggested that complexity can arise through other routes. Some argue that life has a *built-in tendency* to become more complex over time."(Emphasis ours.)

These scientists include Daniel W. McShea and Robert N. Brandon at Duke University, who show that organisms' parts differentiate over time and generate mutations that produce complexity. Useless or deleterious ones die out but others spread through the population and can contribute to selection advantages. They dub this action "the zero force evolutionary law." Just so, Michael Gray of Dalhousie University in Halifax and Joe Thornton of the University of Oregon produced compelling evidence working with very

40. Matthew Fox, *Creativity: Where the Divine and the Human Meet* (New York: Tarcher/Putnam, 2002) pp. 7-8.

complex fungi. They call their process "constructive neutral evolution." Removing genes that would take away some of their versatility, these fungi produced more parts that began to diverge from one another and ended up from this disorder producing more complex structures. They "degenerated their way into complexity."[41]

Summing up his article, Zimmer quotes the highly regarded biochemist David Speijer at the University of Amsterdam: "Speijer says, 'that Gray and his colleagues have done biology a service with their idea of constructive neutral evolution.'" Speijer goes on to say, "Everybody in their right mind would totally agree with it."[42]

George Francis Rayner Ellis, the Emeritus Distinguished Professor of Complex Systems in the Department of Mathematics and Applied Mathematics at the University of Cape Town and co-author of *The Large Scale Structure of Space-Time* with physicist Stephen Hawking, is considered one of the world's leading theorists in cosmology. He has devoted many decades to the study of causality, that is, "Where does the cause of things come from? Does it come from below, adding up the parts from the bottom up, which is the reductionist view believed by most science. Or is it possible that it somehow comes from above, that is, to fulfill some higher purpose, what is called top-down causation?"

Reporting in the *New Scientist*, Dr. Ellis says:

> My ideas developed through conversations with biochemists and philosophers and since then it has become clear to me how ubiquitous and important top-down causation is. It is a counter to strong

41. Daniel W. McShea and Robert N. Brandon, *Biology's First Law* (University of Chicago Press, 2010).
42. D. Speijer, "Making sense of scrambled genomes," *Science* (February 2008) p.319.

> reductionist ideas, which I believe misrepresent the way causation works in the real world. As scientists focus more on the emergence of complexity, taking this into account will become increasingly important. Top-down causation provides a foundation for genuine emergence, where complex systems with new kinds of behavior emerge from combinations of simple ones.[43]

He also points out that "this is the way order arises from disorder."

The major finding of our work is that science itself can now demonstrate and verify that these two ordering and disordering forces make up the cosmic dance, a subtle and beautiful creative dance across the universe. This dance is always in the process of breaking things up and then putting things together in new, different and more unified ways. From disorder it is creating new and higher order—perpetually generating greater unity. This has long been a spiritual concept that is now supported by the rigor of scientific evidence.

Science and spirituality join hands!

The *tendencies* of order overcoming the *tendencies* of disorder results in a continuing creative process. In humans this is referred to as using divergent and convergent thinking: generating many, often nonsensical ideas and then bringing logic in to sort them out and produce something new. This mixing of "disorder" with "order" results in a "creative outcome"—a product, a painting, a poem, a musical score or a smartphone.

In nature it results in what physicists call an "emergence": an atom, a cell, a human being—all resulting in the creative wonders we see in the world of nature.

---

43. George Ellis, "Time to turn cause and effect on their heads," *New Scientist* (August 21, 2013) p. 29.

Again, disorder spawns order. Complementarity acts as a single unitary force.

The noted integrationist thinker and author of *A Theory of Everything*, Ken Wilber, says, "To put it plainly, to say that 'ultimate reality is a unity of opposites' is actually to say that in ultimate reality there are no boundaries. Anywhere."[44]

For us, discovering this interplay of disordering and ordering in the overlay of special relativity and in the conclusions of the second law made in our minds the display of the Fourth of July pale by comparison!

And we found the answer to a gigantic question we did not intend to answer when we began writing this book: What unifies spirituality and science?

## From the Science Side

The principle of progressive bonding redefined the entropy part of the second law of thermodynamics. The two correct answers in the theory of special relativity are active in *Creative Connecting* through bonding in subatomic particles, atoms, molecules, cells, organisms, gene exchanges, and evolution. This proves that the merging of the forces of ordering and disordering continually produce new, novel, and unpredictable complexity—emergences and creations that produce greater and greater wonders and unity.

## From the Spirituality Side

The principle of *Creative Connecting* continually brings greater unity in our lives and in the universe.

Active in *Creative Connecting* are objects, knowledge, ideas, and people. It is a process that continually improves life and our relationships, human to human, and human to all

44. Ken Wilber, *No Boundary: Eastern and Western Approaches to Personal Growth* (Boston: Shambhala Publications, 2001) p.25.

of nature. This profound principle is supported by "love one another," "The Golden Rule," and acts of compassion, kindness, caring, empathy and love and our highest morality and joy.

The identical force of *Creative Connecting* thus unites science and spirituality.

Creative Connecting shows how our lives fit into the greater scheme of things, demonstrating our unity with the stars in the vastness of space, with the most minuscule particles of matter, with all living things and with the sacred and higher power of the universe.

"And over all these virtues put on love, which binds them all together in perfect unity" (Colossians 3:14).

Love is the greatest connector of all.

## The Principle of Syntropy—The Law of Creative Connecting

Nature universally and continually creates more and higher levels of connectivity among things, ideas and peoples. Its purpose leads to total unity in the cosmos.

There is simply no reason to continue to separate science and spirituality, or science and religion, for that matter. Each can learn from the other. Each can gather insights and intuitions that can reveal ever more truth about our wondrous universe.

"The religion of the future will be a cosmic religion. It should transcend personal God and avoid dogma and theology. Covering both the natural and the spiritual, it should be based on a religious sense arising from the experience of all things natural and spiritual as a meaningful unity." Albert Einstein, 1954.[45]

---

45. Fraser Watts and Kevin Dutton, *Why the Science and Religion Dialogue Matters* (West Conshocken, PA: Templeton Foundation Press, 2006), p.118.

Let us be clear. Without having spent most of our lives dealing with creative processes and under the tutelage of the people we mention in our acknowledgements and organizations like the Creative Education Foundation, we would never have had the knowledge or audacity to see the possibilities of universal creativity at work—the unifying that comes about through Creative Connecting.

"And in the case of superior things like stars, we discover a kind of unity in separation. The higher we rise on the scale of being, the easier it is to discern a connection even among things separated by vast distances." Marcus Aurelius, *Meditations.*[46]

That is exactly what we see in nature and in human creativity: the intermingling of order and disorder with the pull of the future. We see the creation of the new and different, the uniting of opposites, the bringing of new unities; winning the ancient tug of war, the universal dance of nature moves *toward greater unity.*

And we are on the universe's team.

As we found in all of our research into human creative thinking, the process always includes the generation of many alternatives: many, many possibilities for solution. Certainly Edison's list of thousands of ideas for a light bulb filament would look like chaos or Linus Pauling's endless notes about vitamin C would decorate an asylum; this is the case with creative thinking.

And so it is with nature. Researchers are often asking, "Why so much variety?" In the chaos following the Big Bang, scientists estimate a cosmos of $10^{87}$ particles—that is, 10 followed by 87 zeros. Out of this, the total number of elementary subatomic particles surviving comes to less than four hun-

46. Marcus Aurelius, *The Emperor's Handbook.* Translated by *C. Scot Hicks and David V. Hicks* (New York: Scribner, 2002), p.105.

dred. And of these only a few were selected to go forward. Out of the vastness of space protons found electrons, a perfect match and an atom—something *new* was formed. Something that goes about getting together with others and creating just exactly the mixtures we would need for life to begin. This play of variety continued. In biological creativity, the Cambrian explosion created thousands of life forms, but only a very few went forward. An array of humanoids walked the earth. *Homo sapiens* alone went forward. Nature experiments endlessly. Just like we do! Quantum or entropic "disorder" feeds the universe with new connections and possibilities. Nature takes that vital step of creating many possibilities!

Without the "disorder" of a blank canvas and the numberless combinations of shape, color and line, how would a painting emerge? Or could we see the mind of a poet and the jumble of words that tumble in chaos before words are set to paper, we would see nature at work.

The universe demonstrates laws of possibilities and probabilities. Nature acts as a creative consciousness. It creates an infinite number of possibilities, and through the history and experiences of the past a few are selected to go forward.

As Nobel Laureate Ilya Prigogine put it in the conclusion of his book, *The End of Certainty*, "What is now emerging is an 'intermediate' description that lies somewhere between the alienating images of a deterministic world and an arbitrary world of pure chance."[47]

Yet it is not just "pure chance," but as Werner Heisenberg put it, "[T]he atoms or elementary particles themselves are not real; they form a world of *potentialities or possibilities* rather than one of things or facts." (Emphasis ours.)

What we can now see is a design beyond beautiful,

47. Ilya Prigogine, *The End of Certainty*. Translated by Odile Jacob (New York: Simon & Schuster, 1996), p.188.

expressing constant creative transformation moving to join everything together. Our spiritual life acts to inform and assist in the constant transformation of the universe in our every act every minute of every day.

We are nature's most evolved transformative agents. In every act of our lives we can ask, "Do my actions contribute to bringing us and our world together or do my actions push us apart?" It is our choice!

Gravity, the universal force, the universe's act of universal love, helped the dispersed matter of the early universe congeal into a progressively connected universe by attracting all things that exist to each other. The other forces in nature, deep within atoms, continue that process of connecting in more powerful and magnificent ways.

None of us would be here if connecting was not the fundamental driving force guiding the universe. Atoms would not form, molecules would not congeal, cells could not merge and life would not appear.

All of this is the beginning of the exquisite expression of the natural creative connecting process that always leads to more connections. What we see is a creative process: having many, many alternatives and selecting from them. This process is repeated over and over and over again throughout the universe. It is a simple, beautiful and profound story.

It not only reveals a universal process, but a grand intelligence that not only created the universe but is constantly recreating the universe. And we are co-creators in this process. Our lives have meaning beyond any we could ever have imagined, from our lives and daily activities to the grand collaborations of many people on the planet.

Who could have imagined that putting a brighter light on and asking a new question about this box experiment and equation could reveal not only the purpose of unification of the universe, but also how each of us can connect our existence and daily lives to the fabric and grand principle of the cosmos?

Yet, that is what it does. It reveals that the most fundamental activity of the universe at all times and everywhere is to creatively connect things, ideas and people.

We can look at anything around us that is "disordering" and recall that as time passes most everything will go on to re-order, regularly at more complex and interdependent levels. And we are getting better at this "recycling" all the time. Your old car becomes your new refrigerator. Tires go from being on the road to being the road itself. Even as things disintegrate, atoms and molecules are freed to reconnect at more and more complex levels. We are reminded of the bumper sticker that says "S—t" happens." We suggest that it should be accompanied by another sticker that says "And grows things."

CHAPTER SIX

# DISCOVERY AND REVOLUTION

---

Scientists were totally shocked by what they found. Finally, by 1925, science had penetrated to the most fundamental level of nature to figure out how nature really worked. What were the laws that nature followed to give us the world and universe we know? With that knowledge, the last secrets of nature would be revealed and endless possibilities for humans to control nature would emerge.

Once more we will see nature's hidden force working.

Scientists found themselves in an extraordinary quandary. *What they found could not be true!* Science makes up a continuing story of discovery: a narrative of more true theories replacing false ones. Up to now each discovery had, in some fashion, built upon past knowledge. Even with relativity, some scientists had suspected that there must be some vast energy locked inside of atoms. What they found in the quantum search however was so bizarre that they tested and tested and tested what they found. Every test verified what they found was true, even though the concepts made no sense at all!

What did they discover?

Scientists were able, at last, to come up with two equally valid ways to see into the deepest levels of nature. They could be looking at a light particle, an electron, a proton, a molecule, or other particles. One view revealed a solid particle; it had real weight, substance, and position in space—a solid reality. The other view, looking at the same thing, revealed nothing there. The particle was *gone*. The only thing there was a mathematical formula that said that whatever was there occupied all of space, weighed nothing, and had no substance at all.

So nature at its most fundamental level was made up of solid particles and nothing but mathematical clues existed to show that there was a "something" there.

This became known as the "wave/particle duality."

Every experiment physicists could come up with showed that this very strange idea about the workings of nature was true. Science simply had to accept what had been found, no matter that it could not be understood. And as we noted earlier, the physicist that is perhaps the most notable expert on the subject, Richard Feynman, concluded, "Nobody understands quantum mechanics."[48] One of the discoverers of quantum mechanics, Niels Bohr, said, "If anybody says he can think about quantum physics without getting giddy, that only shows he has not understood the first thing about them."[49]

Even though nobody could understand quantum physics, there were aspects of the theory that could be applied. Just from a practical standpoint, it is estimated that applications of quantum physics now make up over half of the gross domestic product of the United States. It makes all the electronics around us work: cell phones, TVs, satellites, MRIs, CAT scans, and so on.

Why would we bring up this strange enigma in this book? When you bring up this mysterious subject, even physicists tend to head for the nearest exit. At the beginning of this

---

48. John Farndon, *The World's Greatest Idea* (London: Icon Books, 2010), p.194.
49. Richard Feynman, *The Character of Physical Law* (Cambridge, MA: MIT Press, 2001), p.129.

book we noted that one of the things that had piqued our curiosity was the concept that thinking about something could make it come true; some examples we cited were "The Power of Positive Thinking," the "Law of Attraction" and the like. We mentioned that we had never been able to apply this in our own lives, as the various authors that we had read could never explain *why* this might be true.

We wanted to know the reason, the logic that could explain how thinking about something could affect its outcome.

It turns out that only by looking at physics through the lens of consciousness can the paradox of physics be explained, and only by understanding the laws of physics can the power of our consciousness to help us achieve our goals be understood. They illuminate each other. Quantum mechanics affects our lives in many, many ways, from all the technology that surrounds us to the way we think. It makes up the most fundamental truth of nature! It makes perfect sense as we can now see it through the lens of Creative Connecting.

Why is this new understanding of nature so important for us on our journey through life? Two things stand out in importance about these extremely odd phenomena.

One is to help discover that we are not only made up of the stuff of the universe, but that we already carry the deep knowledge of this strange nature in our innermost being. Second, once we understand what is happening, we can use its power on a daily basis. Each of us carries within us the intuitive knowledge of the most profound meaning of existence.

This chapter in many ways is a reaffirmation that you are indeed a child of the universe and project its image. You naturally carry within you the deep understanding of the forces of a creative cosmos. You know stuff you don't know you know. It will be useful to know you know it.

Quantum physics shows that what and how you think will create your future.

To take full advantage of the quantum discovery, bear with us as we first take a look at the reality of the physics involved. We will stay away from the mathematics and esoteric stuff. We believe it is worthwhile to understand the powerful reality of this "quantum" world so that when you use it, you will know it is the real thing.

Let's review. Quantum mechanics revolves around the fact that everything exists in two completely different and contradictory states at the same time. Take an electron (but it could be a photon, a proton, a neutron, a molecule, etc.). If you look at it one way it is a hard particle. It has mass, weight; you can find it in a particular place. It is a *real* object. If you look at it another way it is what is called a "wave," that is, an invisible "something," an ephemeral emptiness spread out all over the universe that cannot be found in any particular place and weighs nothing. How can this possibly be? Yet it is bedrock fact proven in countless experiments and in the real world.

Quantum physics demolished the traditional, logical, rational, sensible worldview in the world of science—and this theory has gradually spread to the general population. First let's define the term "worldview." A worldview is our generally agreed upon set of assumptions of how the world fits together and works. Our worldview is the means with which we make basic sense out of our world. The new facts from the world of quantum physics give us a seemingly nonsense-based worldview: one that is unpredictable, where things do not happen necessarily because of a cause, but happen somewhat randomly and inexplicably.

Not only do we have a world made up of "somethings and nothings," but when the "somethings" happen, such as particles of reality, they happen *randomly*. Scientists used to believe that events could always be traced to a prior event that was its source—a world unfolding with mathematical certainty, where the future was determined by the past. The new reality of quantum mechanics showed that at nature's deepest level, particles appeared at random—unpredictably: the inescapable lesson of quantum theory. Look for where an electron is and you will only find a probability that it is there, or perhaps somewhere else. The mainstream view of science is that this uncertainty is a fundamental feature of everything. "Some aspects of Nature are, at their very heart, governed by the laws of chance."[50] As we will see, the laws of randomness are the very same laws that fuel creation!

The quantum revolution leaves us with an understanding of the universe that says that everything exists simultaneously in two opposing states: a real material thing and an unreal thing called a "wave," often now called a "field." Now you see it, now you don't. And you never know when or where.

The term "wave" used by science, in this case, is *not* used to designate something like the wave that appears when we drop a pebble in water, but an invisible pattern in space and time that accompanies matter. This kind of wave is more like a crime wave or a wave of selling on Wall Street. A wave in physics is actually a mathematical description of the probable futures of a material particle. All the building blocks that make up our physical world carry in their present state a wave of future potential. They are both being and becoming at the same time. And they are both quite real.

The wave state of an atom carries the potential for neutrons, protons, and electrons to achieve an enormous variety of possibilities. Without the wave state, new creative possibilities could not occur. Atoms would be inert or dead. Atoms would be incapable of growing and connecting. The wave is nature's method of connecting and bringing together atoms and molecules. The wave phenomenon holds our world

---

50. Brian Cox and Jeff Forshaw, *The Quantum Universe: Everything That Can Happen Does Happen* (New York: Penguin Books, 2011) p.7.

together. Without the wave of their potential, atoms would just hit each other and bounce off. They wouldn't be able to merge. Because the waves of an atom can join with other atoms, new possibilities can come into being. This is just like you reaching out and connecting with another person: an entirely new realm of possibilities emerges.

Heisenberg, one of the great physicists and a contributor to the discovery of quantum physics, once described this wave state: "It meant a tendency for something. It was a quantitative version of the old concept of *potentia* in Aristotelian philosophy. It introduced something standing in the middle between the idea of an event and an actual event, a strange kind of physical reality just in the middle between possibility and reality."[51]

Neutrons, electrons, and protons carry a wave that is as *real*, *concrete*, and *important* as the object itself. It's real because the future carried in the wave *is* going to happen. If you don't know that a wave is part of matter, you'll miss the opportunity to use its possibilities. This was a major discovery in unlocking the secret of the atomic and subatomic worlds. Only with this knowledge and the use of the wave reality could the amazing breakthroughs and "magic" of modern science and technology happen. And this applies to your new kind of future as well!

Ok, so we have a situation in quantum mechanics with the probabilistic randomness of uncertainty that is downright weird. But we also have the findings that we prove through experiments that very practical things can be done, though still bedeviled by that randomness, unpredictability thing. There are very useful aspects of this confusing theory.

If the wave/particle discovery was not enough to transform scientists' understanding of nature, there was another totally unexpected result of understanding the workings at

---

51. Des MacHale, *Wisdom* (London: Prion, 2002).

the atomic and subatomic levels. This provides an astonishing illustration of the true forces driving change as we saw when we took another look at special relativity. It's worthwhile to look from a different perspective at this question of "causality." What is it that drives or motivates my actions? As we saw, the old idea has been that we are "determined," that the past moves us—it causes what we do. When the combined concepts of special relativity that we propose are combined with the pull of the future (the pull toward a purpose, connecting at ever more interdependent levels), the future is subtly *guided* but *not* determined by the past.

"What nature demands from us is not a quantum theory or a wave theory; rather, nature demands from us a synthesis of these two views which thus far has exceeded the mental powers of physicists." Albert Einstein.[52]

For a moment, we must look at how physicists study electrons or other subatomic particles. By doing this, we will uncover the "pulling" force propelling change and resolve the enigma and uncover the force of the wave that we can ride in our own lives.[53] Let's engage in a simple experiment that most of us have seen in science classes. We are only reporting one aspect of this very interesting experiment here. The idea here is to observe electrons hitting a target after they have passed through two very tiny openings in a wall and impacted on a target screen.

---

52. Sanjay Moreshwar Wagh and Dilip Abasaheb Deshpand, *Essentials of Physics* , Volume 2 (Dehli: PHI Learning, 2013), p.614.

53. We urge readers to tap the internet and look at the "double slit experiment," particularly something called the "delayed choice" part. There is a YouTube video at https://www.youtube.com/watch?v=DfPeprQ7oGc. See also Ross Rhodes, "Wheeler's Classic Delayed Choice Experiment," at http://www.bottomlayer.com/bottom/basic_delayed_choice.htm. This extensive technical report provides a detailed empirical proof of the future pull we are discussing.

Instead of simply passing through and continuing on a straightforward journey to the target, every time they end up passing through the screen and forming *parallel bands* at the point of impact. This should not happen.

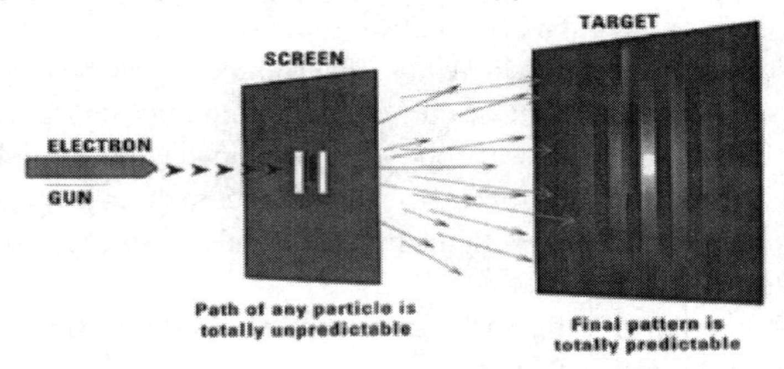


At first, this doesn't make any logical sense. Subatomic particles are like little bullets, so why does the pattern happen? Also, since countless experiments have shown that the path of any single electron passing through the openings is absolutely unpredictable, the final pattern should end up in a random distribution. Yet, mysteriously they somehow collectively "know" what to do and where to go and how they fit together. The same patterns of light and dark bands are always created.

This was finally explained by the idea that the *waves of the particles* are meeting each other, and like waves in water they are either reinforcing each other or diminishing each other and thus producing alternating bands of light and dark. This makes sense and verifies the wave nature of nature.

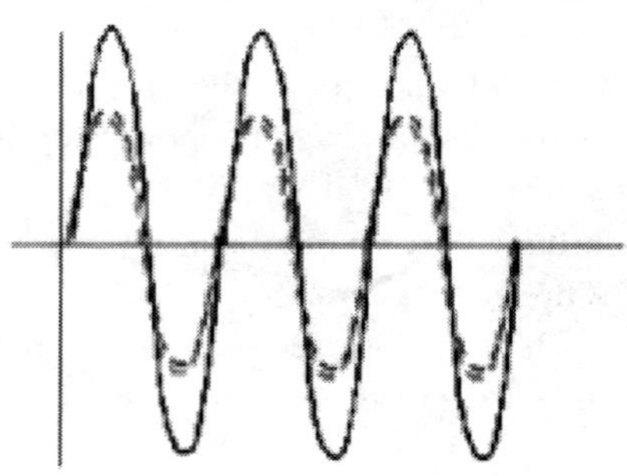

Think about waves for a minute. Imagine you are in a pool. Your buddy splashes a wave toward you and you splash one back. The solid line is his wave and the dotted line is yours. At some point they meet. If they coincide and the high point of his wave meets the high point of your wave they add up and the combined wave gets bigger (higher). Where the dips meet they make a lower dip. If these were light waves, the high points would make brighter lights and just so, the dips would make dimmer lights. These would create the alternating bands we see on the target screen. A whole bunch of light waves go through the slits and like the water waves they *interfere* with each other producing the highs and lows on the screen. So far so good.

But, it doesn't matter whether the electrons are fired at the target rapidly in a bunch or one at a time over a very long period of time so that only one electron is going through the slits at any one time. They could be fired once a minute or once a month. In some extraordinary way they still know where to go to form the same pattern. In no way can the cause or source of this pattern-making of particles be found in their past—or even in their observable material present.

A physicist might try to study the track of one of these atomic particles and make a prediction based on its past direction and the result will always be wrong. On the other hand, predicting the future pattern of many particles is always right, no matter when they are fired. Complete order comes from total disorder! How can randomness and disorder add up to predictable order?

Releasing the particles one at a time gives no chance for electrons to interfere with one another—yet they still make the pattern of light and dark bands! How can this possibly be?

This enigma may appear to be a rather obtuse finding that might interest only scientists, but it carries *profound and far-reaching meaning and power* in all of our individual and collective lives. It has to do with how each of us creates our future, our wave of possibilities, and the results we get. Most of our self-imposed limits come from the assumption that has prevailed for almost all of recorded history, that is, that *the present is a result of past causes.* This automatically leads us into sort of automatically recreating the past instead of being able to fully create the future we want. We have been defeating nature. Let's take a look at how we can change our idea of the future.

Take the activity of a radioactive atom. It is known that such an atom will disintegrate over time, but when each nucleon will leave the atom is completely unpredictable. It has nothing to do with either the atom's past history or its present situation. Some invisible force is acting on it, something far beyond a past or present cause. Since it is unpredictable when any single nucleon of an atom will leave the atom, any reasonable, logical mind could conclude that the atom might occasionally emit most or even all of its nucleons at once. It never does! Astonishingly, however, half of the nucleons will decay precisely in a fixed time frame—every time. This is what gives radioactivity its well-known half-life.

This adding up of large numbers of unpredictable events and ending up with predictable results is what is often referred to as the law of large numbers. Although no logic can explain how it happens, it is very useful. The great mathematician John Von Neumann referred to this bizarre behavior as "black magic." The magic of probabilities, however, works! It allows insurance companies and gambling casinos to turn a regular profit. Statistics show that six-and-a-half words will be added to the English language every day of this year, seventeen out of a hundred ties bought will be for Father's Day, and one out of every eight restaurant bills will be incorrect. It is predicted that one in six teenage girls in a particular city will become pregnant. And that exactly one half of the nucleons in any carbon-11 atom will decay in twenty minutes. Uncertainty transforms into certainty!

The catch is, how do people know how many ties to buy and for what purpose? Who will tell which girls in that city to get pregnant? Do restaurant workers all over the United States get together and decide when to stop adding their bills correctly? How did carbon atoms calculate exactly when to start and stop emitting nucleons? Arthur Koestler felt this phenomenon was due to blind coincidence or we are "driven to opt for some alternative hypothesis."

The resolution of the paradox appears when we shift our viewpoint and search for another explanation. A shift away from past cause to one of *Future Pull* provides just such a concept. It allows the enigmas of the new physics to join as factual background for the unceasing and relentless evolutionary process of life. It, moreover, gives us a new kind of common sense to apply to our turbulently changing world.

The paradox appears only because of the ancient assumption that causes come only from past events. If we step around to the other side of the events and imagine that events can also be caused by the future, top-down causality, the mystery dissolves. If natural events, atomic to human, have a direction, if they are being pulled to the future, then while single events may stray, the total system will always assume a predictable—and interconnected—pattern. As the noted physicist George Ellis points out, "Sometimes the lower level entities only exist because of the nature of the highest level structures. This is the case for all symbiotic relationships, where the partners are unable to survive when separated. They can only exist in the context of the interacting whole."[54]

54. George Ellis, "We Need to Rethink the Hierarchy of Cause and Effect," *New Scientist* (August 17, 2013) pp. 28–29..

This is a pattern and a *force* that we can take advantage of throughout our lives.

A simple thought analogy demonstrates the concept of Future Pull. Take the experiment we began with in which random electrons end up as predictable bands on a target. Imagine taking a box of tiny iron filings and throwing them out at random on the top of a table. What would be the result? A disorderly arrangement naturally happens. Now imagine a large magnet under the table. Again we throw the iron filings out randomly. This time, while each tack will land in its own unique place, the final result will be very different. The bunch of filings will be pulled by the magnet and will automatically be arranged in a predictable and orderly pattern.

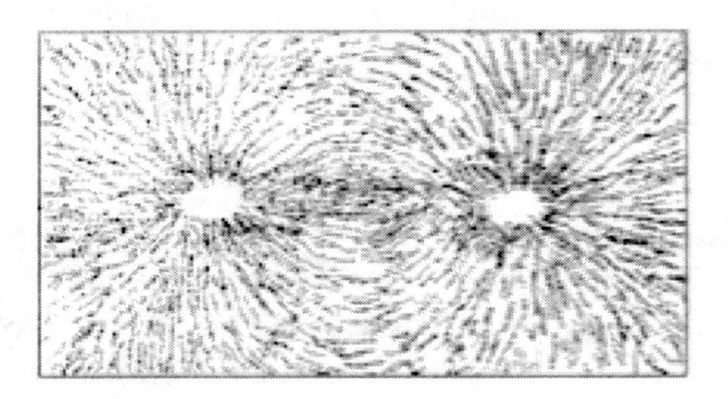

*The pattern of iron filings with a magnet under the table. This pattern will always form.*

While the past surely affects whether the electrons exist in a particular spot in the first place, in the end they act like they were attracted to some kind of a magnet. They are pulled by the invisible forces of the future just as surely as the magnet pulled the filings. Where an individual filing ends up is unknown.

Some people say, "Well, I study market trends or track results to try to predict stock prices or horse races. I'm looking to predict the future. If causes come from the future I can beat the odds."

There is a vast difference between past or historical trends and future pull. Past trends represent specific historical facts. Future pull is a future *tendency*, not anything specific. It is unpredictable as regards specifics.

It is the tendency to pull things together, to bring things closer. It will not predict the specifics of a horse race. It will not predict the position of a particular iron filing.

Let's return to the dots on the screen as they appear when the waves "collapse" to form particles. It is "observed" by the screen or any detector or any observer. It moves as a wave and when observed it collapses into a particle—and simultaneously, it immediately creates a new wave.

When physicists first saw this strange activity in nature they believed that an "observation" collapsed the wave. They concluded that *the observation itself created the reality.* This was indeed so perplexing that they concluded that it would not be productive to examine this quandary further.

Too many discussions ensued to report here but the result was something they called the "Copenhagen Interpretation." It holds that quantum mechanics does not yield a description of an objective reality that we can understand, but it could be called "Complementarity" and deals only with probabilities of observing, or measuring, various aspects of energy quanta (entities which fit neither the classical idea of particles nor the classical idea of waves). Since it works in practice, it is alright even if it makes no sense at all. As we noted above, one of the discoverers of quantum physics, Niels Bohr, put it, "If anybody says he can think about quantum physics without getting giddy, that only shows he has not understood the first thing about them." It may be strange but just do the calculations. They work. They give us a language to talk about and manipulate nature and that's good enough.

Now we can understand this seemingly strange phenomenon.

How could it be that just observing or "measuring" something could change it? It turned out that it was about interacting with something that did change it, but in the meantime the notion spawned all sorts of foolish ideas: just thinking or envisioning something could create it. Envision a Cadillac and somehow it would manifest in reality. This idea has pervaded all sorts of semi-philosophical and, as the British put it, "fruitloopery" idea systems. But somehow, in some cases this concept seemed to actually work! The real question is why?

This was very confusing to many famous Nobel physicists like John Wheeler: "If you are not completely confused by quantum mechanics, you do not understand it."[55] And yet another Nobelist, Schrödinger, also said about quantum mechanics, "I am sorry I ever had anything to do with it."[56]

Why such an enigma? It seems like a paradox because the world is made up of protons, atoms, electrons, etc. that are real objects with mass, weight, and that occupy a definite position in space. But they also exist as a "wave," not a real wave as such, but a mathematical description that has wave-like properties that spreads all over space. It cannot be pinned down to one place, has no mass and disappears when it encounters something else, sort of like a ghost. How could nature be made up of something and nothing at the same time? And when you looked at large numbers of events of these objects, they took on probabilistic patterns that could not be predicted by their past.

How does this seeming paradox really work?

This can be explained—by looking at consciousness!

---

55. Vishal Sahni, *Quantum Computing* (New Dehli: Tata McGraw-Hill 2007) p. 18..

56. "A Quantum Sampler," *New York Times* (December 26, 2005) [www.nytimes.com/2005/12/26/science/27eins_side.html]. .

We actually experience this phenomenon every minute of every day. We as human beings are always experiencing the *now*, the particle "being"; it is always accompanied in our minds by our continually thinking about the possibilities of the *future*, the wave of "becoming." These two states of thinking happen simultaneously all the time: the reality of this moment overlapping with the ephemeral, invisible possibilities that lay ahead.

This happens in the core brain system as shown in the research done by Daniel L. Schacter, Donna Rose Addis and Randy L. Buckner at the Department of Psychology at Harvard University. What research shows is an integration of the present with past memories and future possibilities: an overlay of the disorder of the multiple possibilities of the future with the guidance of the known order of the past, a guidance toward what we hold in our minds as a goal.

And what about the resolution of the great quantum paradox? You and everything else in the universe exist *simultaneously* in two apparently contradictory states. You exist in the real now, your being, your "particleness." You are a real object, if you will. You have mass, position, size, shape, etc. You also simultaneously have your "possible" future, your becoming, your "waveness," which spreads out all over space and actually is not quite infinite. Just as quantum physics finds, it is probabilistic: influenced by the history of your past that narrows your choices but does not determine them. The future is real: one of those possibilities in your mind *will* happen, just as a wave "collapses" in quantum language. Becoming shifts to being.

This is all just common sense—and creative intelligence, just like the creative intelligence of the universe. You have the concrete reality of the present and the "wave" of your possibilities for the future. And, to repeat, just as was found in quantum physics, that wave is not infinite; it is influenced by your experiences of the past so you end up with probabilities, not infinities. As the great physicist Feynman found, you have to consider "sum over past histories." What he means by this is that we must consider the momentum of the past as it influences the future—and not just the past that happened but also the past that probably could have happened.

The minute you do something, your *now* changes, the multitude of future possibilities "collapse" into what you just now did. You have a new *now* (particle) and new *future probabilities* (wave) and just exactly like what happened in quantum physics, the old possibilities "collapsed." You are a living, breathing expert in the interplay of the paradox of wave/particle physics.

Ok, right now you are sitting there in your comfortable chair, it's Friday night, and the future is looming. You have a ticket to tonight's ball game, one that Harry, your best friend, has been begging for. Your wife wants to try that new restaurant or go to see a movie. Niggling at you is also that vacation on the island in the Caribbean you have been promising your wife and you have the vacation days coming. You could sweep her off her feet and catch a plane for a few days of fun. Wow, there are *almost* infinite futures out there! (Although a few are more probable than others.) Welcome to the wave world.

Why is that important, in your life or in anyone's life?

The universe is always *pulling* toward *more connecting*, just like gravity is always pulling things together. That is the direction of the force—uniting. That is the magnet. If your movement flows toward helping form more or better connections—be they simple connections of joining something better with a nail or repairing a road or perhaps something like getting people together—you are connecting. Or you could even connect better with your spouse. You could be taking advantage of the pull of the future. You could be in tune with the universe. Connecting.

Maybe that new car you were dreaming about will be used to deliver Meals on Wheels. You could be riding the wave! Or as they say in Star Wars, "May the Force be with you."

But, if you are just doing things for only your own benefit or breaking things up—not so you can put them together better in some way, but just to destroy or disorder them—watch out.

Ask yourself, is what I do adding or subtracting from our connectedness, our unity?

What are you carrying in your "wave?" What is your vision of the future?

Beyond the connecting pull of the future, just as in the physical universe, your wave is also *real.*

In the physical world of subatomic particles, electrons, protons, neutrons etc., the wave carries a future determined by the universe. It provides the tendency, the positive force,[57] for particles to join in more complex configurations and form molecules, cells, plants, animals and lead to humans and to *consciousness.*

Consciousness is totally different! It makes it possible for us to envision the future. It makes it possible to *deliberately form the wave*, not simply be driven by default by the basic forces of the universe! Of course we can simply be carried along with the fundamental forces—or we can choose to create our own wave, a vision, of our own: a reality that we can help to make true.

Just imagine this however. You are presented with thousands upon thousands of stimuli every day. How does your brain know which one to pay attention to? If we have an image of our future we have created something like a radio antenna. The brain has a way of sorting through all that vast

57. In the quantum physics equation used to solve the wave function (the Schrödinger equation), this *lowest energy* wave function is real and positive definite - meaning the wave function can increase and decrease, but is positive for all positions. It physically cannot be negative.

multitude of stimuli and integrating the information that is in alignment with your image. We will automatically become aware of the most important information that will help us achieve our goals. If we are carrying negative thoughts or just coasting we are subject to what is called "inattentional blindness" and we will likely miss important information.

This helps answer the question that had bothered us for so long. There is a reason for the success of "positive thinking" or "the vision thing." There is a deep and scientific reality in how we think about our future. In the next chapter we will talk about the new research that shows how we affect our wave reality.

Is our vision appreciative and positive? Do we hold it as real? Is it in alignment with the purpose of the universe – of more and better connections?

## CHAPTER SEVEN

# FUTURE PULL

---

We have reached a critical point in our journey together: recognition of the most vital quality of the change process is connecting. Connecting is not only constant but change has a direction; it is pulled by the future.

In his book, *Chance and Necessity*, Nobel Laureate biologist Jacques Monod stated, "One of the fundamental characteristics common to all living beings without exception is that of being objects endowed with a purpose."

Just so, when two people join to create a child, although the past is represented in their genes, the future is based on the unique way those genes combine. That wave of potential pulls the trillions of cells forward to self-organize into a unique human being. A hybrid plant pulls a flower or fruit—never before known—into being. In exactly the same way, the resonance of the waves in a particle of quartz pulls it forward into a crystal.

What is that future? Seeming miracles, such as the emergence of our continually evolving planet, the creation of life, or the appearance of human intelligence, are but commonplace and continual creative activities of nature's world. The concrete scientific evidence daily accumulating in the study of life, evolution, the development of civilization and our own thinking processes shows a world progressing inevitably toward more complex relationships and deeper connections. This could only happen if change had a purpose.

Or as José Ortega y Gasset puts it:

*Life is a series of collisions with the future;*
*it is not a sum of what we have been,*
*but what we yearn to be.*

Change is driven by the pull of the future to connect everything at broader, deeper, more interpenetrating levels.

The pull of the future manifests itself in our lives, our dreams, our visions, our hopes, our waves of possible futures. Look around you, most everything you see and take for granted is an outcome of someone's idea of a possible future. Often those ideas were considered to be crazy or idealistic. "Wouldn't it be wonderful," someone once thought, "if a house could have holes in the walls you could see through and still be protected from the weather?" "How about having heat or cool air whenever you want it?" No power is greater than ideas and ideals, the waves that continually create the future.

Of course, as one of the very successful folks we talked to emphasized, "Yes, you have to have a purpose, a vision but if you don't work your a** off you'll never get anywhere!"

We have not realized until now the *reality* we create with our visions. Yet, as we look at the people or organizations enjoying great success, we find in every case it was no accident. They allowed the pull of the future to work—and the push of their work—to propel them far beyond the limits and boundaries of the past. Their wave of the future acts like a radio receiver; they tune into the world around them and connect with anything that fits their vision. The past is not pushing them: they allow the powerful reality of Future Pull to work.

As we have seen, with the advent of human beings and consciousness everything changed.

Atoms, electrons, photons, protons, etc. depend 100 percent on the ubiquitous force of the universe to connect and form more connected and complex arrangements. The change with the mind is that we are no longer dependent on the built-in force of the universe. *Consciousness gives us the opportunity to shape the wave that is moving us.* This explains the success of such concepts as "the power of positive thinking" and the "Law of Attraction" that have brought so much success to so many. The fact is that in our minds we can envision the future we want. We can shape the wave that is moving us, not just depending on the universe or our environment but on our own vision of the future. We will "tune in" to those clues in our world that connect with that wave and not with some neutral or negative vision. We will automatically make choices that take us in the direction of our dreams. The wave *is* a reality that is happening, the becoming that our being and our energy is moving toward.

The principle of Future Pull poses critical questions for each of us entering our future. What is pulling me? What is my purpose? What is pulling my organization? Success and deep personal fulfillment come when one is fully open to connecting with those individuals, circumstances, and events that will lead to the fulfillment of one's purpose. The challenge is to put the future in the driver's seat by harnessing the energy and commitment of your life to the power of Future Pull.

The universe and we share a grand purpose.

In our culture, we believe that goal setting, hard work, and discipline all join together to achieve anything worthwhile. We give little thought to the power of doing what you love to do. Once you have found your natural creative gifts, the accomplishment of great goals becomes much easier. You can extend the phenomenal force of Future Pull into your life by asking where you are going. What is your purpose? What vision do you have for your life?

You are the only one who can decide what makes you happy. If you don't know what makes you happy, then you really have to stop and ask yourself why. What stands in the way of your knowing yourself well enough to know what really makes your heart sing?

A compelling purpose energizes life. It is the force that activates our "wave of becomingness." It brings forth happiness and joy. Without a compelling purpose we live life as a fairly haphazard experience, being easily swayed by the latest fad, temporary pressures, or the most recent advice on what others think we ought to be doing with our lives.

George Bernard Shaw believed the true joy in life was "being used for a purpose recognized by yourself as a mighty one; the being thoroughly worn out before you are thrown on the scrap heap; the being a force of nature instead of a feverish selfish little clod of ailments and grievances complaining that the world will not devote itself to making you happy."

Having a purpose allows us to be creatively pulled to the future. We are able to build on the past while escaping its bondage! From limits and expectations, we move to possibilities and potentials.

Outstanding leaders have known the potent force of an inspiring vision. Deep commitment to a great purpose joins the organization and each individual within it to the vitality and energy of the natural pull of the future.

"You don't have to be a genius or a visionary or even a college graduate to be successful. You just need a framework and a dream." **Michael Dell**, creator of Dell Computer.

**Mary Kay Ash**, founder of Mary Kay Cosmetics, had a dream that she could make cosmetics that women would love, make a profit, and treat the people who sold her products with respect and concern. She created a multimillion-dollar success.

**Steve Wozniak** and **Steve Jobs**, founders of Apple Computer, imagined changing the world by giving the average person access to the world's knowledge through personal computers.

**James Rouse**, the great urban planner, had a vision to improve urban life along with giving his people gratification for the quality of their work. Faneuil Hall Marketplace in Boston, South Street Seaport in New York City, and Baltimore's Harborplace, completely altered these urban environments for the better and attest to his compelling vision.

Legions of outstanding successes in business demonstrate the seemingly extraordinary force of Future Pull. Even in huge enterprises the natural power and central thrust of a compelling vision bring all of the different aspects of the corporation into a coherent, vital, and energized whole.

Most organizations have no vision. Complex multimillion dollar enterprises operate with a limited mission statement as their compass to the future. "Our mission is to return maximum profitability to our shareholders." "Our mission is the long-term increase of shareholder equity." "Our mission is to achieve 15 percent profit before taxes." Is it any wonder that most people within these organizations are uninspired? There is no forceful purpose, no picture of the future whole, no internal DNA for self-creation or even to provide a reference point for decisions. Most of all, the loss of the real and powerful natural force of Future Pull occurs when a compelling vision is never created or shared within the organization.

A 57-year-old chief operating officer of a huge military industrial company called together three hundred of his top managers. Citing lagging profits and high costs, he told them, "The only purpose of this company is to make a top return on the capital invested. We're not around just to build the best products." Little did he know that many of his top managers would leave within a year and that poor morale would sink even lower.

The former president of Notre Dame, **Father Hesburgh**, feels that "The very essence of leadership is you have to have a vision. It's got to be a vision you articulate clearly and forcefully on every occasion. You can't blow an uncertain trumpet."

Individuals face the same challenge as organizations. They must find what they deeply care about, what gives their life passionate meaning and then harness those things to a compelling purpose. Each of us has been gifted with our own remarkable way of expressing our humanity. Recognizing this is our challenge.

**Debbie Meier** is trying to do nothing less than create a new system of public education, and she's doing it in the most unlikely place—Harlem, New York. She took on this impossible task of creating a new school in Harlem to try out her theories of education. She founded the small school movement. "Largely my ideas about teaching and learning focus on democratic values, by which I mean a respect for diversity, a respect for the possibilities of what every person is capable of, a respect for another person's point of view, a respect for considerable intellectual rigor."

**Anita Roddick** had a $6,000 bank loan and an idea she loved: a business that would sell body care products and that would care deeply about nature, about employees, and about customers. She founded The Body Shop. Fifteen years later, with no advertising, with an organization with sales over $150 million, her electricity and passion still infuse the enterprise. She says of the people she works with, "You want them to feel that they're doing something important, that they're not a lone voice, that they are the most powerful, potent people on the planet." Her commitment runs deep. Customer service, employee education, and concern for the environment matter—success and profits naturally follow.

The Canadian-born **Guy Laliberté** began his circus career eating fire on the streets, playing accordion, walking on stilts and juggling. He brought his team and dream to the Los Angeles Arts Festival in 1987, *without a return ticket*. Result:

the world famous Cirque du Soleil we know today. Today, Laliberté is worth $2.5 billion.

**John Paul DeJoria's** parents divorced when he was two, and he sold Christmas cards and newspapers to help support his family before he turned ten. He was eventually sent to live in a foster home in Los Angeles. He was a L.A. gang member before he joined the military. He worked for Redken Laboratories and got a $700 loan to create Paul Mitchell Systems, holding on to his dream of a quality product and living out of his car. Now JPM Systems is worth over $900 million annually.

**Leonardo Del Vecchio** grew up in an orphanage and went to work in an auto parts and eyeglass frame factory. His vision was to have his own factory. Well, Ray-Bans and Oakleys now have six thousand retail shops like Sunglass Hut and LensCrafters. His estimated net worth is now above $10 billion.

**Ingvar Kamprad** lived on a farm growing up. But he always had a vision for business, starting by buying matches in bulk and selling to his neighbors. He later expanded to fish, Christmas decorations, and pens. He expanded to mail order and got local manufacturers to keep his prices low. It worked! It became IKEA.

**J. K. Rowling** lived on welfare and wrote books. She created the Harry Potter franchise.

He milked cows and sold magazines in Oklahoma, and imagined rural folks could somehow get good value at low prices; then **Sam Walton** founded Wal-Mart.

Wearing dresses made out of potato sacks, she ran away from home at age 13. Finally, her mother sent her to live with her father. She got a scholarship to college and won a beauty pageant and she turned hardship into vision. **Oprah Winfrey** created a multibillion-dollar empire.

**Steve Binder**, a public defender, had a vision to move homeless people charged with misdemeanor offenses away from the courthouse and toward community involvement.

Binder's program resolved 3,700 cases in one year and has inspired more than 35 similar courts across the country. "People on the street want to live their lives more fully and lawfully," he says. "They need an opportunity to do so."

**Ali Brown** said: "I trusted myself. This was my path. I listened to my heart and knew that this was what I was supposed to do." Starting with $18.67, she founded Ali International, a dynamic enterprise that is devoted to empowering women entrepreneurs around the world, and currently has over 65 thousand members in her online and offline programs.

**Maxine Clark** had crazy visions indeed! Maxine took an idea for custom-made bears and has now turned Build-A-Bear Workshops into one of the world's most successful toy companies.

Great visions will not take you anywhere if they are not accompanied by equal dedication and energy and, as Theodore Roosevelt put it, "Stick to it" enough. It takes full dedication to climb over the many obstacles in your path—but that's what a great vision can drive you through.

Your purpose not only inspires and gives meaning to your life but makes nature your silent partner in the evolutionary journey. The most powerful purpose comes out of giving your unique skills and talents to make the world a better place. From teacher to carpenter, from manager to musician, we all have special gifts to give. To uncover your own purpose, focus on those unique qualities that describe who you are, and then look at the ways you enjoy expressing them. What do you love to do? What makes you happy? What is it you dream of becoming? What do you dream of giving to others? These are the questions that will lead you to your purpose.

Joseph Campbell calls the commitment to a purpose following your bliss. "If you follow your bliss, you put yourself on a kind of track that has been there all the while, waiting for you, and the life that you ought to be living is the one you are living. Wherever you are—if you are following your bliss, you are enjoying the refreshment that lies within you, all the time."[58]

Today, it is estimated that of all the people that go to college only 15 percent ever work in the field of their major. And, within that huge college population, only two percent of them will do what they really love to do with their lives.

Because money has become the badge of success, particularly in the United States, the brightest young people move away from their primary creativeness. This reality was brought home to us as we watched two fine young men make their career choices. One loved filmmaking, the other was an accomplished musician and was gifted at playing a myriad of musical instruments. They were both in college and their parents insisted that if they pursued their dreams of doing with their lives what they loved to do, they wouldn't ever be able to make a living. Today, both of these young men are headed for law school. One still plays his music as a hobby and the filmmaker has quashed his dream. Now, what kind of lawyers are they going to make?

Selecting a life's path based on secure employment and the money it will garner is hardly going to lead to finding what one loves, one's real identity, or what one can truly pursue with passion. These early choices lead many to end up miserable and unhappy, without pride in their work, without the drive to excel—because they aren't doing what they love to do.

We have not listed the very many who have achieved their vision that was not marked by the rewards of money. They are hard to identify. But there are scores more of them than those that hit the monetary mark. You probably know some of them. They are the happy ones, doing what they love.

---

58. Joseph Campbell, *The Power of Myth* (New York: Anchor Books, 1991)..

Lynne McTaggart, in her extensive survey of the research on consciousness, reported, "Human consciousness is increasing the order of the rest of the world and has an incredible power to heal ourselves and the world: in a certain sense we make the world as such, as we wish." She added, "Living consciousness somehow is the influence that turns the possibility of something into something real. The most essential ingredient in creating our universe is the consciousness that observes it."[59]

If you still harbor doubts about how your mind can affect the future, the fact is that actual experiments have now been made that have tested this concept in the laboratory. The question was, "Can the mental state of observers affect the wave state of anything? Can it change reality?"

Let us consider the science of the wave/particle quandary. As we pointed out with the double slit experiment, the result of the pattern formed on the screen comes about from firing an electron particle through a slit. Its wave passes through the slit, interacting with other waves and affecting where the wave "collapses" into a new particle on the screen. We can observe that there is a hidden wave this way. These waves lie outside the spectrum that can easily be detected, like radio waves and such. Without this kind of experiment nature's hidden force would not be revealed, as these waves are quite invisible. But there is much more here. What is the "reality" potential of that invisible force?

What did the experimenters find when they had observers watching the experiment?

> [T]he scientists found that the very presence of the "detector-observer" near one of the openings caused

59. Lynne McTaggart, *The Field: The Quest for the Secret Force of the Universe* (Harper Perennial, 2008).

> changes in the *interference pattern* of the electron waves passing through the openings of the barrier. ...*Thus, by controlling the properties of the quantum observer the scientists managed to control the extent of its influence on the electrons' behavior.* The theoretical basis for this phenomenon was developed several years ago by a number of physicists, including Dr. Adi Stern and Prof. Yoseph Imry of the Weizmann Institute of Science.[60] (Emphasis ours.)

This is quite unlike the phenomenon we experienced in the early days of television where fiddling with the TV antenna or just moving around it would cause the TV to flicker. These wavelengths and interference patterns are not affected by physical objects moving around in their vicinity.

You might want to read this again. In plain English it states that the mental state—the "properties"—of the "observer" affects the physical reality of the wave—what the physical future of the event will be! The experiment shows that our thinking can actually have some real power over the reality of the future.

In terms of the possibilities this represents, consider this:

> You get an electron with infinite mass, infinite energy, and infinite charge. There is no way to get rid of the infinity using valid mathematics, so, the theorists simply divide infinity by infinity and get whatever result the guys in the lab say the mass, energy, and charge should be. Even fudging the math, the other results of QED (quantum electro dynamics) are so powerful that most physicists ignore the

60. *Ibid.*

> infinity and use the theory anyway. As Paul Dirac, who was one of the physicists who published quantum equations before Schrödinger, said, "Sensible mathematics involves neglecting a quantity when it turns out to be small—not neglecting it just because it is infinitely great and you do not want it!"[61]

Your vision, your picture of your future will guide you and not only "tune you in to your goals" but physically, in some way, even affect the reality of the future! Is yours a vision of hope or one of fear?

A "Quantum Reality" of now, of being—and the one of the future in your mind, of becoming—is what makes up your life.

Your "becoming" is in your hands, in your mind—in your conscious creation of the future waves that will carry you to your future. The next chapter will show some other examples of nature's hidden force.

The "Law of Creative Connecting" is creating more, broader and deeper connections among all things in the universe. The basic force and spirit driving the universe starts with gravity, the greatest connector, and continues this process in everything, all matter and all life.

The universe is *being pulled to the future*, moving inexorably toward more and deeper connections—and, as we will see in the next chapter, doing this through a dynamic *creative* process.

Connecting, connecting . . .

Our language, justly, has invented many ways to express connecting:

---

61. John Gribbin, *In Search of Schrödinger's Cat* (New York: Bantam Books, 1984) 257-259.

**connect**: affix, ally, associate, attach, bridge, cohere, come aboard, conjoin, consociate, correlate, couple, equate, fasten, get into, hitch on, hook on, hook up, interface, join, join up with, marry, meld with, network with, plug into, relate, slap on, span, tack on, tag, tag on, tie in, tie in with, unite, wed, yoke, accompany, add, adhere, agglutinate, annex, append, assemble, blend, bracket, cement, clamp, clasp, clip, coadunate, coalesce, combine, compound, concrete, conjugate, copulate, entwine, fasten, fuse, grapple, hitch on, incorporate, interlace, intermix, juxtapose, knit, leash, lock, lump together, mate, melt, mix, pair, put together, splice, stick together, tie, tie up, touch, weave, weld, pertain, affect, appertain, apply, ascribe, assign, be joined with, be relevant to, bear upon, compare, concern, consociate, coordinate, correspond to, credit, have reference to, have to do with, identify with, impute, interconnect, interrelate, orient, orientate, refer, touch, appear with, belong to, characterize, co-occur, coexist, coincide with, come with, complete, follow, go together, happen with, occur with, supplement, take place with, hitch to, nail, screw to, integrate . . . and so on.

Look at your own life.

Think about what you do—even what you think are the most mundane activities. You fix dinner; you connect with the ingredients at your grocery store. When you get home you perform destructive connecting: slicing, chopping, making a mess. Then you blend (connect) things in a new order so the food will connect with your family. At work you put things together: words, knowledge, ideas, and people. You might fill potholes in the road; you might design a house or figure out which medicines you should connect with a patient. You're in the connecting business—part of the universal conspiracy to unite everything.

One day we were visiting a client, a very successful one, and stopped for a moment to chat with a janitor who was cleaning the entrance. "What are you doing?" we asked. "Well," he answered, "I'm keeping the place clean." "It looks pretty clean already," we said. He responded with a big smile, "We want our customers to know that we care about them and keep a nice place for them. I make that happen." He was connecting and felt some real pride and joy in what he was doing.

Beyond the evidence from such fields as cosmology and biology, in the end, the reality test of Creative Connecting is in your life and mind. How does it agree with your ordinary, everyday experience? What is the source of natural pleasure and pain? How do we feel about the world and our lives?

With very few exceptions, nature has provided an invaluable internal compass: as Epicurus suggested thousands of years ago, *connecting gives us pleasure, disconnecting yields pain.* As we experience a connection of concepts, knowledge, ideas or people it triggers a sense of gratification. Just so, in being denied or experiencing a disconnection we can feel a range of negative emotions. Losing a loved one produces the greatest pain of all.

Beyond that, nature provides us with our values and ethics. Even at the atomic level nature's success depends on bonding, connecting, in mutually beneficial ways. Without what is called covalent bonding we would not be here. Gravity itself lies at the foundation, the ultimate "DNA" of life. Mutual sharing is the key.

Gravity connects everything in our cosmos. It is totally non-judgmental; it does not decide that it doesn't like one plant or atom and not connect with them. Gravity never judges, nor does gravity refuse to connect. All things in the universe are held together by gravity. Gravity is the ultimate expression of unconditional love.

The eminent philosopher William James believed that truth is a quality the value of which is confirmed by its effectiveness when applying concepts to actual practice (thus, it is "pragmatic"). Truth is verifiable to the extent that thoughts and statements correspond with actual things. Truth "hangs together," or coheres, fits as pieces of a puzzle might fit together, and these are in turn verified by the observed results of the application of an idea to actual practice. James said that "all true processes must lead to the face of directly verifying sensible experiences somewhere." He also extended his pragmatic theory well beyond the scope of scientific verifiability, and even into the realm of the mystical: "On pragmatic principles, if the hypothesis of God works satisfactorily in the widest sense of the word, then it is 'true.'"

Our issue is simply to look closely at how the universe behaves—what it does.

And in the next chapter, we'll see what fantastic shifts it produced in the past and what it is doing today. What great surprises it has in store for us now!

CHAPTER EIGHT

# TODAY'S AND YESTERDAY'S BREAKPOINT WORLDS

We have talked about the understanding of entropy, the move to syntropy, the Law of Creative Connecting, and many of its implications. But what about the tumultuous world we face today? How does Syntropy apply to the mess we're in: financial crises, climate change, wars, terrorism, energy shortages, refugees, uncertainty, unemployment, political stalemates, poverty—we could go on and on. What to do? Where do we look for guidance? Einstein said, "We cannot solve our problems with the same thinking we used when we created them." As much as we respect Einstein this is not much help. Where can we look for a different kind of thinking?

The universe could be such a source. The universe is pretty smart.

We can look at an amazing shift nature made that changed our entire planet. It is very similar to what we face today. Nature applied Creative Connecting in an extremely significant way to make the biggest transformation this planet has ever seen. If you lived at that time and studied the history of the earth you would be sure that life was headed to sure destruction. There was *no* way that anything learned in the past would provide a solution to the crises that life faced.

A billion years ago, if the old law of entropy was all that nature had up its sleeve, life could not have survived. There were not enough natural resources. The climate was changing radically. Competition was forcing whole massive colonies to perish. Any economist would predict that crowding alone would push life past its limits. Nature had worked very hard to produce life and it was about to disappear!

Not only did it not die out but it came up with a completely new, completely unpredictable, extraordinarily effective *creative* solution. It applied syntropy.

How did life get to this point? And what was the solution?

The solution we can apply today?

We will go back in time to see how life painted itself into such a corner and how we have done the same thing.

Looking at the over thirteen billion years the universe has been in business we can see that from the "Big Bang" a vast gas cloud, made up of very simple atoms and molecules, shifted as the universe acted to gradually clump these tiny molecules together in larger and larger masses and connected everything to everything else in the universe. Over thousands of millions of years, cosmic dust connected to bring into being countless constellations. Some of these clumps became stars; others became planets, asteroids or comets.

In our case, after about ten billion years the earth and our solar system came into being. From a dead rock, a billion years of fiery volcanoes, bombardment by asteroids, extreme climate and atmospheric changes, little clumps of atomic groups joined together to form an incalculable myriad of mixed molecules.

One of the many atoms involved in this tumultuous stirring ferment of possibilities was carbon. Carbon is the connector beyond any other atom. Carbon can make an enormous variety of bonds with itself and with other atoms.

Life of our kind exists in the universe only because the carbon atom possesses certain very exceptional properties.

After about a billion years of experimenting, about three and a half billion years ago, exploring countless combinations in a harsh and chaotic environment on earth, nature hit on a special formula for connecting the environment. A very large complex molecule was organized around carbon. A connecting factory was invented—"genes." This special molecule informed atoms how to transform the environment into countless little copies of itself—"living" cells, much like today's bacteria.

Nowhere has nature created anything remotely as efficient as these early cells. A typical bacterium can grow and divide every half hour or so, so in one day one bacteria cell can produce about 16 million daughter cells. Under the right conditions, some bacteria can divide once every 20 minutes. If there were no other factors involved, a single bacterium could divide 80 times in a day.

This would lead to approximately $2^{80}$ bacterial cells, that is 1,208,925,819,614,629,200,000,000 cells. Over two billion years you get a lot of cells. Talk about smart! Nature could turn rocks, water and sunlight into life!

Gigantic communities of living cells transformed the earth. Given the variety of environments available, as far as we can see today, it would appear molecules combined in numerous ways in different kinds of communities with different attributes, from some in arid, hot or temperate environments to others thriving in seawater and freshwater lakes.

While these early cells were amazingly efficient, in time they faced a huge problem regarding their ability to continue connecting in their surroundings. While they transformed the earth over several billion years, *they themselves did not change*. Each community was made up of practically identical cells. Replication became the standard. In the face of radically changing environments, enormous climate change and the problem of such effective and rapid growth, these cells faced crowding and competition among different communities and dwindling resources. Such little change occurred

in these cells themselves over eons that biologists call these times the "boring billion(s)."

A seemingly insurmountable problem presented itself. In a changing environment new threats emerged. Stuck as they were in a single successful pattern, these early cells *could not adapt*. They suffered "replicative vulnerability." In a community of identical copies, if a disease or pollution or resource depletion or an effective competitor occurred or a predator threatened even one, it would affect all! Entire and vast communities could be wiped out. What to do? If continuing to make more and deeper connections makes up the direction of nature, how does growth, making more and deeper connections, continue? After all, nature had over two billion years of great success with the *duplication* paradigm. A very new kind of growth solution was needed!

Suddenly, sex reared its head!

Here again is another example of nature's creative connecting process at work. It selects for those big changes that could continue making more and deeper connections with each other and with the environment.

Let's look at the implications of this next soaring cellular leap to our own life and times.

A billion years ago this giant leap of cells sharing genes profoundly re-created our world.

It's happening again! We live in a radically changing world. Our world is being re-created.

What we are talking about is the whole nature of change itself: what it is, where it comes from, how we can control it, and at its deepest level, what the process of connecting and change really means for each of us individually and collectively. When we say that things are "changing," what does that really mean? Have we made some small adjustments or is something actually transforming in a major way? A huge spread of possibilities emerges when we talk about change. In the following pages we will discover that *change itself actually changes* and what worked at one point can actually be counterproductive at another point.

What concerns us in the body of this book is the *direction* of change and how it affects our lives and what it *means* in the overall scheme of our lives.

Just like those cells a billion years ago we are now facing the advent of a new *kind* of change. Relentless revolutions are engulfing our planet. Everything is changing turbulently, unpredictably and relentlessly. We are facing challenges that we have never seen before. It is paramount that we find new ways to understand the changes reshaping our lives. We need a clear understanding of what is causing these changes and what we can do about them. How can we understand this kind of unprecedented and unpredictable change? We need some guidance to lead us into a future—a future unlike any encountered before.

We believe that nature has provided this kind of guidance. After all, as we have seen, nature has been in the business of dealing with big changes for over thirteen billion years.

Let's look at that big change.

For several billion years success had been built on the basis of separate types of cells competing for resources and fighting to keep their particular identity and maintain their growth—growing their colonies larger and larger. This method was starting to fail as different cell colonies competing for the same resources created diminishing returns and battles for territory.

The solution for all these problems came "out of the blue."

It's all about sex. Wouldn't you know?

Sex equals mutualism, and gene sharing was created.

Different kinds of cells would get together, share capabilities, find ways to explore and grow in new environments and *both* would benefit. A plus B would make C, an unpredictable organism with many more and different capabilities.

In an amazingly clever way nature created a totally new way to move ahead. It shifted to the most gigantic and dramatic change that has ever been seen on our planet—sexual reproduction, what is known today as genetic recombination.

Out of the innumerable cells created, some of these simple organisms "merged" to form a super intelligent organism. After experimenting over two billion years, a billion years ago simple cells formed mutualistic symbiotic teams that shared their genes and evolved into a single organism—an organism that could produce an enormous *variety* of offspring. The result of this union was the first "eukaryotic cells" — the type of cell that makes up the human body. All the plants and animals and we humans owe our very existence to these tiny organisms that connected in an amazing new way, not only combining genes, but incorporating other cells into their bodies in collaborative partnerships!

By combining genes in myriad ways, the diversity created among its offspring population now assured that some of the new and different cells could counter the new threats and proceed to occupy a much wider variety of environments.

These new cells were and are still amazing. They operate in a brilliantly systematic fashion. The remarkable molecular systems that underlie this organization make up one of the most astounding processes in molecular chemistry and biology.

Here again was an example of the creative process at work: always creating an immense variety of possibilities and selecting for those that could continue making more and deeper connections with each other and with the environment.

Entirely new and different organisms were created with new gene combinations.

Life shifted to sharing and cooperation!

Mutualism evolved along with gene sharing. For example,

fungi needed more oxygen than they could get from the soil, and trees needed more minerals than they could get from their root system. They made a partnership. Trees would feed the fungi from the air and fungi would feed the minerals from the soil to the trees. They could both thrive in places that previously could not support either. This ancient collaboration continues today.

This is nature's creativity at work. Nature combines mutually exclusive things to bring about a third capability that can do something new and solve a problem.

Unpredictability comes to the fore again and again in new studies of evolution. Until recently a long-held assumption regarding the appearance of new species was that a new species would appear as a result of an accumulation of small changes that led to the result of a creature, plant, etc. that could not reproduce with a prior species. This is the characteristic of a new species. This assumption could not be tested rigorously until recently due to the lack of DNA data.

For the first time an abundance of cheap and speedy DNA sequencing became available to evaluate this old idea. Mark Pagel, an evolutionary biologist, and his collaborators at the University of Reading in the U.K. were able to set up a study with a large amount of data from trees, cats, bumblebees, hawks, roses and the like to see if the evolutionary DNA branches showed accumulations of small changes or some other manner of species change.

To their surprise small changes *did not* lead to new species–only six to eight percent of new species occurred as a result of these small changes. Seventy-eight percent were brought about by unpredictable single changes. They were *not* driven by environmental pressure. As Pagel puts it, "It isn't the accumulation of events that causes a speciation, it's single rare events falling out of the sky, so to speak. Speciation becomes an arbitrary, happy accident when one of these events happens."

Just as we see in other cases of "emergence" and "creativity," the rare, the novel, the unpredictable happen and new creative connections drive evolution.

"The overall collection of technologies bootstraps itself upward from the few to the many and from the simple to the complex. We can say that technology creates itself out of itself. This mechanism can be called evolution by combination, or more succinctly, combinatorial evolution."[62]

How many people decried the idea of government getting involved in business when they "bailed out" the auto industry? Government had no business in business! We now see the result. This is just one example of "organisms" that are "different" collaborating to bring about a new solution.

We live in a world and have a history of success gained through separation, duplication and competition. This is like nature's very old idea of cellular success.

Today we face a "clash of cultures" where each culture refuses to acknowledge any good in its opposition as they compete for the world's resources and continually end up in a "lose-lose" result. This is the way it has always been: rigid rules, separation, boundaries and competition. Simply put, if nature chose to continue to work this way, human beings would never have come about on this planet.

Possible changes to come include:

A move from competition to cooperation.

A move from separation to collaboration.

A shift from more of the same to brand new answers.

Business, government and finance collaborating to create best possible solutions.

Governments might create businesses (as in China).

Schools could educate in innovation and make money in royalties from their inventions.

---

62. W. Brian Arthur, *The Nature of Technology: What It Is and How It Evolves* (Free Press, 2009) p.13.

Let's get together and create the future!

This is not some kind of utopian dream. Like most of our technology, all we have to do is copy nature. Nature practices creativity, sharing, mutualism, collaboration and cooperation. If you cut your toe, the whole body responds. Right away. It does not have to go through departmental approvals.

Think just a minute about nature's creative intelligence. Your body contains trillions of cells. Somewhere in you, it is thought—your mind—that helps all these cells work together to give you a healthy body. Just imagine that! Imagine such a human organization. Unimaginable! All working together! A plant growing in your yard works at a higher level of intelligent organization than the best company on the planet. Not only that, it can make electricity in ways we still don't understand!

Nature's Law of Creative Connecting will help us uncover the forces driving this change: how we can make today and tomorrow's change an ally and not an enemy. If we understand the forces driving change, our future can actually be extraordinarily better than we have ever imagined.

Just like those early duplicating cells, our old solutions that have worked so well for so long are simply not working in this new global environment. In many cases the old solutions are even making things worse. Because so many organizations have grown like those early cells, by making copies of themselves, they suffer the vulnerability of replication. When the environment changes radically and organisms or organizations cannot quickly adapt, any threat affects *all*, not just a few.

With new electronic communications our planet is creating a new nervous system. When everything is changing radically, understanding the Law of Creative Connecting can help us understand and deal with these extraordinary times. A completely new view of change can turn the problems of our changing world into new possibilities to make our lives better.

In 1922 Pierre Teilhard de Chardin proposed that something he called "the noosphere" would emerge through the interaction of human minds. As humankind organizes itself in more complex social networks, the more the noosphere will grow in awareness. He saw this as an extension of Chardin's Law of Complexity/Consciousness. The law argued that the noosphere is growing towards an even greater integration and unification of mankind, culminating in what he called the "Omega Point," which he saw as the goal of history: an apex of thought and consciousness.

The purpose of Creative Connecting leads in the same direction that Pierre Teilhard de Chardin proposed—a great union. Who knows how long this may take? Or what it might look like? How many of us could have predicted the World Wide Web just a few years ago? Change is accelerating every day. It is worthwhile to remember that natural creative processes most often begin with disorder, a disorder that makes way for the new, the different, and the unpredictable higher order.

If this kind of "super optimistic" view seems a bit Pollyannaish remember this: a billion years ago all of life was powerfully threatened. After two billion years of success, life was failing. Logically, how did cells transcend this much history of success? Resources and the limited capability to utilize them would ultimately cause massive environmental collapse.

Taking all that long history into consideration, who would expect that life could soar into a far better way of transforming itself and its environment? But it did!

Nature continually makes life better, continually transforming and creating higher, more inclusive forms of Creative Connecting.

How do we accomplish nature's Creative Connecting process?

- By tapping into our latent creative capabilities, we can imagine and share new and different solutions that far transcend the answers from the past. Cells tapped into that latent capability by sharing their different "ideas" in the form of genes, which then in these new combinations make up new "solutions."
- By recognizing that a new "nervous system" is bringing us into more and more connectedness among things, peoples, knowledge and ideas. Just as cells shared neurons and adapted to local conditions and made new connections, we can make new organizations and technologies that bring people together.
- By creating new bonds with people who are "different" and combining into collaborative relationships—like different types of cells such as fungi, for instance, combined with algae and a multitude of other mutualistic interactions throughout nature where differences provide the enrichment that nourishes creativity.

We are informed by nature again and again about the creative process. When the new sexually reproducing cells gathered together in colonies over the next five hundred million years, they ultimately joined together into multicellular plants and animals. Many of these organisms were pretty crazy, quite unlike anything that has appeared before or since, resembling dirt-filled bags, flat discs, or even quilts. As with any creative process, the fact is that the "Cambrian explosion" produced an astonishing variety and diversity of living multicellular creatures, only a very few of which continued their evolutionary journey.

"Imagination is more important than knowledge," said Einstein—and this applies to all of nature, we might add.

All creative processes produce many alternatives, with only a very few being selected—those that *connect* best with each other and their environment. This happens even at the sub-atomic level where so many "non-rational" particles exist (hundreds). Thus it is referred to as the "zoo."

The creative diversity produced through sexual reproduction was accompanied by another form of connecting—*collaboration.* As the noted naturalist John Muir put it, "When I go to look at something by itself I find it is hitched to everything around it." Ecosystems are extremely complex with many forms of symbiotic connections. Three forms of symbiosis exist: parasitism—where one benefits to the detriment of another; commensalism—where one benefits with no effect on the other; and mutualism—where both benefit from the partnership. Although it is not referred to as "mutualism," the basis of life—the carbon atom—depends on "covalent bonds," which share energy on both sides of the connection and create many weird and wonderful combinations.

As plants emerged in the evolutionary ferment again, not only do we find enormous diversity that is weeded out leaving only the "most fit," but we also always see collaboration. Almost all land plants depend on the collaboration and connections between the plants and specialized fungi. The plant feeds the fungi with carbon compounds that the plant produced from capturing sunlight and the fungi feeds the plant with minerals and other useful compounds from the environment.

A wonderful example of this kind of collaboration can be found in ever-present lichen: a partnership of algae and fungi. Lichens are amazing and thrive in some of the most extreme environments on earth: arctic tundra, rocky coastlines, dry, hot deserts, and even toxic dumps. They are also abundant as air plants on limbs in rain forests and temperate woodlands, on bare rock, and on the surface of exposed soil.

Lichens express mutualism: neither algae nor fungi can grow and reproduce on their own in the hostile environments they can inhabit with their symbiotic partner. The lichen make partnerships that use photosynthesis to manufacture food.

This lichen example illustrates how nature continually works by putting A and B together to create C—something new and different that is much more successful. All this makes up a fundamental process of nature. When different genes combine through sex, new and different possibilities emerge. Arthur Koestler said that A + B makes C is a definition of creativity: what he calls "bisociation." And you can't predict, logically, what will result from these combinations. It serves us well as a definition as we think about our way ahead and our new, creative possibilities.

Bacteria and us? Bacteria live in our intestines along with many other creatures. We can't digest all of the food that we eat. The tiny microorganisms eat the food that we cannot digest and they partially digest it, allowing us to finish the job. The microbes benefit by getting their food, and we benefit by being able to digest the food we eat.

Recall that business of "replicative vulnerability"? This is when a past successful pattern does not or cannot adjust to a radically changing environment. When one fails, like dominos, all suffer. Early cells were copies, so failures spread throughout the life systems and mutualism and gene recombination needed to emerge so life could continue to evolve.

Our world contains many environments and cultures yet we have exported and replicated and standardized old patterns throughout the globe in these different environments. Business systems, educational systems, banking and financial systems, and governmental systems all tend to take on identical forms, so as our environment transforms radically, our traditional systems often do not adjust; many tend to fail. Our challenge is to create solutions far different from the past. And nature has shown the way. We must change our way of thinking—from repeating to creating!

Again let's remember what Albert Einstein said: "The world will not evolve beyond its present state of crisis with the same thinking that created it."

What did Einstein mean?

We think this shift from repeating the ways of the past to creating new and different—better—solutions may be what he had in mind. That is what Einstein did in his thinking. What lies beyond the age-old thinking process of reason and replication? Are we up to the challenge of learning from the creative genius of nature? Can we do what nature did in creating life? Can we do what nature did a billion years ago in re-creating life with sexual reproduction: to go from automatically copying the past to creating unique solutions?

Can we creatively change our thinking to combine different traditions, ideas and people to create new solutions? It will require a new kind of leadership to build new bridges of understanding to renew our planet and move beyond deep seated prejudices and conflicts. It is up to each one of us to consciously participate in this evolutionary leap in human thinking.

In 2004, the Learning Innovation and Technology Consortium convened the CEO Forum on Education and Technology. They identified creativity as an *essential skill*, and suggested that educators find ways to encourage creativity, curiosity, inventive thinking, and risk-taking in students. Innovation draws on creativity, but not as an end in itself. Innovation is about applying creativity to solving problems. They concluded:

> Often, we only recognize true innovation in hindsight—when we see the breakthrough that engendered other ideas or acted as a catalyst for new

> products or processes. So, what does LITC mean by innovation? True innovation can be identified by both the process and the results, so we define it this way: Innovation means the process of thinking and acting creatively to solve an identified problem with the outcome being a new process or product that acts as a catalyst for new cycles of development. Thus, innovation inspires more action—not simply more ideas, but more innovative approaches to putting ideas to use. Innovation thus increases potential and opportunity, and sparks new cycles of thinking—revolutionizing how we learn, how we live, and how we work.

It will require building new alliances and partnering with those with whom we might disagree. The challenge is to be nonjudgmental and inclusive. This leap in human consciousness is calling for a new kind of leader, a leader who is dedicated to finding not just common ground, but to rising to new creative solutions, not just compromises.

"A society's competitive advantage will come not from how well its schools teach the multiplication and periodic table, but from how well they stimulate imagination and creativity."[63]

We must recognize that not all new possibilities will work. We must be willing to "fail fast" and continue to move ahead when something does not work and try something else. This requires goodwill on all sides.

Barbara Marx Hubbard, President of the Foundation for Conscious Evolution, after more than a half century of intense studies of the future, summed it up, "Once again a new world

---

63. Albert Einstein, Letter to Vivienne Anderson, May 12, 1953, AEA 60-716. http://www.asl-associates.com/einsteinquotes.htm

view is arising ... This idea is the culmination of all human history. It holds the promise of fulfilling the great aspirations of the past and heralds the advent of the next phase of our evolution. It is the idea of conscious evolution."

Julian Huxley, the biologist who by and large is thought of as the founder and coiner of the term "transhumanism," discussed this next phase of human evolution in an article written in 1957:

> Up till now human life has generally been, as Hobbes described it, 'nasty, brutish and short'; the great majority of human beings (if they have not already died young) have been afflicted with misery ... we can justifiably hold the belief that these lands of possibility exist, and that the present limitations and miserable frustrations of our existence could be in large measure surmounted... The human species can, if it wishes, transcend itself—not just sporadically, an individual here in one way, an individual there in another way, but in its entirety, as humanity.[64]

It is up to each of us. We must press forward for a move toward conscious creative thinking in all areas of society.

64. Julian Huxley, *New Bottles for New Wine*, (London: Chatto & Windus, 1957) pp. 13-17. http://www.ne-plus-ultra.org/huxley.htm

## CHAPTER NINE

# WHY, WHY, WHY?

---

*"Imagination is more important than knowledge. For knowledge is limited to all we now know and understand, while imagination embraces the entire world, and all there ever will be to know and understand."*

— Albert Einstein

This simple question of "why" has driven the work of the authors for over five decades. This is also the constant curious cry of young children. "Why is the sky blue? Why do some cows have horns?" These kinds of questions can drive adults crazy. Our asking questions one after another has driven some of our family, friends and colleagues to exasperation. It was our unlimited curiosity that brought us together along with our mutual questioning of authority and our shared viewpoint that looking at parts rather than wholes doesn't make good sense. Even though one of us was raised in a small rural farming town and the other in a number of large cities, we are forever asking more and more questions and—fortunately—seeing things from quite different perspectives and "childlike" approaches. It was, after all, a child who noticed that the emperor had no clothes.

These disparate viewpoints have served us well, in that many years ago we began investigating the subject of curiosity, creative thinking and imagination. Collaborating with early pioneers in probing the nature of creativity,[65] we found

65. Sid Parnes, Cal Taylor, Paul Torrence, J. P. Guilford, and John Gowan.

early on that it was actually possible to help people leap beyond reason and unleash their imaginations. This led to working with groups who faced very difficult problems, helping them think more creatively.

We have to admit that our early investigations into creative thinking were very shocking to us. It changed our lives! We had been brought up in a world of psychology that assumed that creative genius was rare. It resided only in those very special people, the Michelangelos, the Newtons, the da Vincis, the Einsteins, the great artists and scientists who led the way for others to follow. The vast majority of people were "normal, average" and just needed to be taught what the great thinkers had laid down.

Some psychologists were not satisfied with these assumptions—in particular, Dr. J. P. Guilford of the University of Southern California. Guilford was able to identify a number of different kinds of intelligence, including creativity, expressed in different ways. What absolutely amazed us is that when we began to apply these tests of creativity, particularly with children, we found that the assumptions we had carried so long were completely wrong. Later in this chapter we'll go into some detail. What really floored us was that adults were not able to express that same level of creativity and imagination as we saw in children. When we went back and retested those same adults later in some cases—where they had changed to very different environments—suddenly their results on creativity and imagination tests soared! How could this be?

In a nutshell, after following the research over years we continued to be astonished! It turned out that the idea that creative genius is rare is totally wrong. It is actually rare for people *not* to have a brain that can function at a high level of imagination! As you read on you will see why we became so excited and why we began to imagine what could be in our world if only our culture could learn how to apply education methods to help everybody tap into this hidden and latent creative ability.

We have to agree with Dr. Guilford's thinking:

> "Psychology should be the chief basic science upon which the practices of education depend. It should have supplied education with the information it needs concerning the processes of understanding, learning, and thinking, among other things. One of the difficulties has been that such theory as has been developed has been based primarily upon studies of behavior of rats and pigeons. As someone has said, some of the theory thus developed has been an insult even to the rat."[66]

David Shenk covered the story of latent creativity in his book, *The Genius in All of Us: Why Everything You've Been Told About Genetics, Talent, and IQ is Wrong.*[67]

Over the years, we found business receptive to awakening creativity and we worked with an extremely broad range of people from illiterate workers to CEOs.[68] We worked on challenges, all the way from inventing a new chair or shampoo, to teams working on the most vexing issues in biology, medicine and quantum physics, to name but a few. As teams dug more and more deeply into their difficult problems it provided us with an exceptional education that crossed over the boundaries separating many sciences and disciplines.

---

66. J.P. Guilford, *The Nature of Human Intelligence* (New York: McGraw-Hill, 1967).
67. David Shenk, *The Genius in All of Us: Why Everything You've Been Told About Genetics, Talent, and IQ is Wrong* (New York: Doubleday, 2010).
68. See Appendix Five.

What we want to do in this chapter is to share with you one of the most basic findings about creative thinking and how the universe shapes our connecting process. We found very early on that there was a deceptively simple and fundamental process involved in deliberately developing new, different and creative solutions.

The italicized items below require *Divergent Thinking*: unbridled judgment-free imagination.

G – Goal – A clearly stated purpose (connectivity)
*Identify Possible Goals – Then Select YOUR Vital Goal*

P – *Possibilities – A very broad creation of possible solutions*

S – Selecting and applying the solution or solutions that best solve the problem (connecting things better)
*Generating the factors that affect how well your ideas will work*
Testing your ideas against the important factors

I – Implementing – *Coming up with ways to carry out the selected ideas*
Putting the selected ideas to work

F – Feedback

As there is a good deal of material available regarding the "hows" of creative thinking methods we will not discuss it in detail here.[69] They range from free lateral thinking to analogical/metaphorical leaps.

What became most vital in our work was helping the members of the groups we worked expand their inborn imagination to encompass a very broad range of possible answers. The italicized items above demand judgement-free imagination: day-dreaming, if you will. This was most often the criti-

69. See the Creative Education Foundation website at http://www.creativeeducationfoundation.org/

cal missing link in solving problems in new ways. We found it possible to help individuals and teams multiply the number and extent of ideas including the "crazy" ones. This was accomplished simply by learning how to turn off the logical-rational part of the brain so the creative part could function at its peak. It turns out that most breakthrough solutions started out as pretty crazy ideas in the first place. It was vital and essential to produce a large quantity of ideas *before* turning to evaluate them.

The nature of this step, creating possibilities that never existed before, requires that we think way beyond logic. We venture into a realm of fantasy and flights of imagination that always seem—at first sight—illogical, unreasonable and unsound, bringing into being what has not existed before—as nature does.

The great poet William Blake could see into nature: "Some see nature all ridicule and deformity ... and some scarce see nature at all. But to the eyes of the man of imagination, nature is imagination itself."

The absolutely amazing thing about quantum mechanics is it is so much like great creative thinking in the *fact* of its randomness and unpredictability that it led to where we are: into discovering that people's minds are not the only place where imagination resides.

It revealed that nature is a universe of *perpetual imagination*, continually creating new and different possibilities. If it were not for the unpredictability and novelty of those weird quantum jumps we would be stuck with a static, unchanging world and a universe that would soon wither and die.[70]

If it were not for the particle, the "natural selection" side of quantum mechanics that sorts out and rejects the unten-

---

70. We wish to acknowledge the prescient work of Richard G. Colling in his book *Random Designer: Created from Chaos to Connect with the Creator* (Browning Press, 2004).

able possibilities, we would drown in an inundation of novelties. With people it is called sorting out among possibilities—making decisions, choosing.

This oscillation—this *complementarity* of processes from divergence to convergence—makes up the Creative Connecting process of continual evolution. This back and forth dance ranges from the physics of atoms to the minds that create innovations and inventions—imagining many possibilities and then choosing from among them.

As Alfred North Whitehead put it, "Almost all really new ideas have a certain aspect of foolishness when they are just produced"—just like we would say about the "jumps" of quantum mechanics. The pioneering investigator of creativity Dr. Sidney Parnes, founder of the Creative Education Foundation, says, "Most descriptions of creative behavior include the opposites, *i.e.*, playfulness and seriousness, fantasy and reality, nonsense and purpose, irrational and rational."

Thomas Edison, for example, one of the greatest innovators in history, actually tried out six thousand possible solutions before he hit on the one that worked to create a light bulb. "I'll try anything—I'll even try Limburger cheese," said Thomas Edison.[71] He ended up with a solution that kick-started the illumination revolution—a piece of burned cotton thread. Dr. Linus Pauling, who received two Nobel prizes, admitted that the one significant thing to do in solving a difficult problem was create countless ideas without regard to whether they made sense. As the famous cosmologist Carl Sagan put it, "Imagination will often carry us to worlds that never were. But without it we go nowhere."

Why would people call on some kind of consultant to spur their imaginations? People were not coming up with new and different ideas. In the late 1960s we were called upon

71. Karl Albrecht, *Brain Power* (New York: Fireside, 1992) p. 43.

to develop a test for NASA that could help them to identify the scientists and engineers that were most imaginative in order to assign them to the most difficult problems. The test we finally worked out functioned very well and someone on our team suggested that since it was such a simple test that we give it to children and find out how they did on the test.

With the support of the U.S. Office of Economic Opportunity, we pulled together a random sample of 1,600 children between four and five years of age.

We were all astonished at the results of the creativity tests. Ninety-eight percent of the children at age five scored in the very high range, what we identified as the "genius" level of ability in imagination. We decided to turn it into a longitudinal test. So, we tested the same children five years later. The exact same test was given. At age ten the very highest scores dropped to only thirty percent with the same test population.

We gave the very same test five years later to the same children at age fifteen. The test scores dropped to twelve percent of the group. The research project ended at that stage. We think because most of the test administrators got depressed.

We know from testing thousands of adults that only about two percent of the adult population, at an average age of thirty, is expressing a "genius" level in their imaginative ability. We have learned well how to suppress human creativity; we have learned not to "rock the boat."

Newsweek reported that Kyung Hee Kim at the College of William & Mary discovered this, after analyzing almost three hundred thousand Torrance creativity scores of children and adults. Kim found that creativity scores have consistently trended very downward. "It's very clear, and the decrease is very significant," Kim says. It is the scores of younger children in America, from kindergarten through sixth grade, for whom the decline is "most serious."

The central questions for us had been whether imaginative thinking is learned or is it genetic and where did their creativity go? And can we get it back?

The answer, as we can all see, is that imagination is not learned; imagination is unlearned. Bucky Fuller once said, "All children are born geniuses and society de-geniuses them." Or as Einstein commented, "It is a miracle that any curiosity survives formal education." He also concluded, "The only thing that interferes with my learning is my education."

We could speculate endlessly as to why we might have designed an educational system to restrain creativity, from economic to sociological factors. The fact is that most school systems around the world produce learning *what* to think rather than *how* to think. We are taught to continually judge and critique everything, the very mental activity that inhibits imaginative thinking. Unlearning this perpetual evaluation activity of the brain and being able to turn it off at will is one of the most critical ways to release our latent creative capabilities. We can tap into that five-year-old's mind that still lives inside us.

Jean Houston, a noted leader in the Human Potential Movement and initiator of The Foundation for Mind Research, after working with thousands of people in all walks of life, from street people to the White House, concluded, "We all have the extraordinary coded within us, waiting to be released." If anyone had doubts about this, watching it happen before your eyes with Jean will remove them forever.

Thanks to nature's grand design, we *never lose* the capacity to regain our five-year-old genius. We can recapture it rather easily. Our research and experience have demonstrated that almost everyone can learn how to tap into their imaginative genius. It is never lost—only suppressed. The inborn ability to create new and different solutions and possibilities never goes away. Creativity is actually part of normal brain activities. The part of the brain that is used to create the marvelous innovations and inventions we now need to make life better for everyone is exercised every day by practically every human being, through the process of dreaming.

We see that particular parts of the brain in the upper cortex get very excited during dreaming. These regions of the brain also get very excited during the exercise of imagination. Yet, during the resolution of logical, rational problems, those areas of the brain show minimal activity. The parts of the brain associated with creative thinking are not needed to remember, or find, the one right, correct answer, so we do not regularly call on our imagination. But that imaginative part of the brain is available whenever the conditions are right.

We can get it back whenever we want!

It is tremendously exciting to know the latent capabilities within each one of us. A compelling key to a new understanding of nature lies in this wondrous and miraculous gift of imagination. The most important thing we want to establish in this chapter is to tap our imagination and learn to work with many possibilities (as children do)—in fact with any problem there are probably almost infinite possibilities. It was our work with creative thinking that unveiled the natural connecting process described in this book.

While many think the physical world is limited, all we have to do is recognize that nature has created all the wonders of the world around us, starting with rocks, water and sunlight. Nature applied its imagination and creativity. There are no limits!

Imagination has always been central to genius throughout history. From Pythagoras to Newton to Einstein, imagination created numberless wonders, from medicines to microwaves, from cell phones to the internet. Just as imagination was necessary for nature to invent new ways to evolve, imagination is absolutely critical in our world of great challenges.

We are surely up to the challenges of our time if we can only learn from nature—from the wisdom of nature.

If we stop believing the current worldview of diminishing returns, hard work, and more worries—the "commonsense" view—and instead take advantage of what we have learned about how everyone can think creatively, we can *all* participate in changing our world and solving the big problems that face our changing world!

But who will lead us beyond our current mechanistic way of thinking and logical, rational way of solving problems – the way we have become comfortable with for so very long? This is our real challenge. To help people change their beliefs about the nature of their minds and education and the possibilities they carry in their own minds, we need persuasion of a very high order. Something dramatic that touches what people already know but has been suppressed. Everybody remembers their five-year-old self and wonders it they went. It is not easy to make such a change. When we ask a group of people for those who feel they are creative to raise their hands very few respond. But, with the right exposure, two hours later all respond! We will not go into detail here. There are abundant resources available about people and organizations that help people tap into their creativity.[72]

We *can* change our way of thinking!

Practically everyone can recapture their childlike imagination and open-mindedness. We can replicate this natural ingenuity worldwide. Indeed, these creative connections, nature-driven successes, suggest a heartening truth. Just like the single-celled organisms that, faced with catastrophe, shifted to sexual reproduction a billion years ago, we humans

---

72. See The Creative Education Foundation, http://www.creativeeducationfoundation.org/. See also Jean Houston, *The Wizard of Us: Transformational Lessons from Oz* (Atria Books, 2012) and Gloria Gannaway, *The Transforming Mind* (Praeger, 1994).

can also tap our latent creativity and creatively collaborate across many "dividing lines," from organizational departments to cultural and social boundaries, and create the kind of miracles that nature did.

We can successfully overcome the great, often thought to be insurmountable, challenges of our modern world. Creativity, resources and will are all we need.

CHAPTER TEN

# UNITING SPIRIT AND SCIENCE

---

*The soul given to each of us is moved by the same living spirit that moves the universe.*

— Albert Einstein[73]

We are nearing the end of our journey together, a journey completely unanticipated when we began this venture. Starting with the goal of finding the source that could be driving creative behavior, we ended up looking at two sides of a giant and ephemeral coin: science and spirituality. Science, on one side, represents the solid, empirical, observable, verifiable world. Spirituality, on the other side, represents issues of the heart, of soul, meaning and mystery, domains greatly affecting creativity. Tangible and intangible worlds, yet surely both essential to the expression of creativity.

We started on what we believed to be the "easy" side—science. But as you readers have seen, a simple pause for a brew at a pub dropped a dead end into our investigation. We were introduced to the theory of entropy, the second law of thermodynamics: a "supreme" natural law that determined that the universe and everything in it was proceeding downhill into greater and greater states of disorder. The law put our human experience of creativity into an odd and unnatural

73. Quoted in *Life Magazine*, "Death of a Genius," May 2, 1955.

corner of the universe—that surely one day we would pay for it when this law caught up with us.

A universe running downhill has been accepted through the law of entropy for a century and a half and many scientists subscribe to such a conclusion. In the latter half of the twentieth century we have often been faced with a *pessimistic* overview of the world: a complicated world, beset with conflict and teetering on self-destruction, a world where the spiritual and the scientific are disconnected and, for all practical purposes, unable to speak to each other.

We could not accept this law! We had always been "maverick" scientists, always asking the awkward questions. When we would ask about the apparent orderliness of our planet, our scientist colleagues would give us such answers as, "Well, our order is being paid for in some other area of the universe with great disorder and it will all even out some day." This did not satisfy us so we, as you have seen, looked more deeply.

Entropy, after all, remained rock solid even after the giant revolutions of relativity and quantum physics and, as you have seen, deserved a second look.

Having discovered that the universe was actually progressing toward higher and higher order automatically brought up new questions, the big one being "Why do things happen at all?"

Over the centuries, three dominant beliefs of *causality* emerged:

- **Greek organic**—Everything had its own intrinsic nature. Rocks fall because they are rocks; everything has its own nature that causes its behavior.
- **Classical physics**—Everything is preordained (determinism). Effects follow causes. If you know all the facts of the past you can predict the future.

- **Quantum mechanics**—Everything is random and probabilistic (quantum indeterminacy). Its causes are unexplained.

The discovery of quantum mechanics has baffled everybody but has won over the physics community with its vast body of evidence and its enormous practical applications—from transistors to lasers to MRIs.

Quantum physics had proven beyond any doubt that the future was just "a roll of the dice," essentially "probabilistic" and *not* deterministic. This was the source of Einstein's famous remark, "God does not play dice with the universe."

This question of causality, the reason for anything to happen, was a big one for Einstein in his theory of relativity. It led us to his work.

The beauty we found in relativity was the hidden causality in Einstein's equations and the possibility of the two-answer solution. It revealed that by combining the randomness of quantum mechanics and the order of determinism a creative process emerged. And this led to the wave/particle duality.

Quantum mechanics revealed that all of physical reality exists in two opposing states; one is a real, material, substance. The other is a "wave," a non-material, mathematical description of possible future states with no physical presence. This was considered a paradox.

The beauty of what was found was described by Niels Bohr, a leading founder of quantum mechanics. "Consequently, evidence obtained under different experimental conditions cannot be comprehended within a single picture, but must be regarded as complementary in the sense that only the totality of the phenomena exhausts the possible information about the objects."[74]

---

74. Paul Arthur Schipp, *Albert Einstein: Philosopher-Scientist* (Cambridge University Press, 1949).

This is, in other words, exactly what we find in the creative process. Two elements *complement* each other: the ephemeral broad randomness (divergence) of ideas combines with solid objectivity and reality of selection (convergence) – seeming opposites.

Scientists have been urged to accept "complementarity" as a natural principle, and now we can see that it is indeed a profound natural principle and the root of nature's creativity.

As we look across the billions of years of history of the universe, this is what we have seen over and over in our observations of nature's creativity. As we have mentioned previously, the use of mathematics precludes results that produce the new, different, unpredictable, wholes greater than the sum of the parts. But nature's creative process moved relentlessly ahead over thirteen and a half billion years. Let's review just a few of those critical, those amazing, unpredictable leaps that defy logic and led to us:

- Particles—new connections among elementary subatomic particles, including photons, in the early universe.
- Dust, asteroids and stars—new connections occur with matter as particles collect and stars originate, mature, and terminate.
- Galaxies—new connections are experienced by stars and galaxies, either localized connections among myriad stars or mergers, acquisitions and juxtapositions among neighboring galaxies.
- Planets—new connections in the physical and chemical properties of planets during the course of their histories.
- Chemicals—new pre-biological connections that transformed simple atoms and molecules into the

more complex chemicals needed for the origin of life.

- Biological evolution—continually forging new connections among life forms, from generation to generation, throughout the history of life on earth.
- Human and social evolution—continuous development of new methods of connecting.

Paul Davies is a theoretical physicist and cosmologist who also works in astrobiology, a new field of research that seeks to understand the origin and evolution of life. Although we and Dr. Davies approached the question common to both science and spirituality from different directions we arrived at the same conclusion.

The noted biologist Stuart Kauffman was very prescient in the conclusion of his book, *The Origins of Order*: "In short, physics is beginning to discover ways in which very complex systems nevertheless exhibit remarkable order. No reflective biologist can view these developments without wondering whether the origins of order in nonliving systems augur new insights for the origins of order in living ones as well."[75]

Sir James Jeans reflected, "In general the universe seems to me to be nearer to a great thought than to a great machine."[76]

The great gift to humans is that of self-consciousness: a consciousness that can ask about the universe, about good and evil. What we have seen in these pages is a universe that we describe as "Creative Connecting."

Nobel Laureate Ilya Prigogine said in 1997, "Our belief is that our own age can be seen as one of a quest for a new type of unity in our vision of the world, and that science must play an important role in defining this new coherence."[77]

---

75. Stuart Kauffman, *The Origins of Order* (Cambridge University Press, 1993).
76. James Jeans, *The Mysterious Universe* (Cambridge University Press, 1930).
77. Ilya Prigogine, *The End of Certainty* (Free Press, 1997).

This syntropy process of continual connecting and unification incorporates two vital forces: one of disordering and one of reordering at higher levels of unity. In the balance of forces, reordering holds the major sway as evolution from star dust to human societies continues. But still disordering (what we often call evil) continues. The pain and suffering go on. This is where consciousness comes into play. We have the power to choose which side to play on.

We have the capacity to identify whether our actions contribute to creative reordering or non-creative disordering. For example, if a nation behaves in a way that is destructive to the world unity, what do we do? One possibility is to isolate them, cut them off from the family of nations so that the pain will force them to change their behavior. Another possibility would be to follow natural law and think creatively and find ways to engage them and find ways to constructively connect with them. Just so, if a person misbehaves, we can disconnect and isolate them away from society or we might find creative ways to bring them into, and connect them with, society. We sometimes call the latter "rehabilitation" when in fact they often never have been "habilitated" in the first place. The consciousness question we face in our decisions is, "Are we acting as creative connecting agents or as disconnecting agents?"

Science and spirituality would tell us we should now be aware of how we are acting as evolutionary agents.

Now science has given us a new and stable platform for making the hard decisions.

In any case, we are practicing the Creative Connecting process, doing the universe's work. We are continuing the vast work of connecting everything in deeper, more interdependent ways.

As we have seen in the last chapter, the capabilities for creative thought are given to all of us at birth. We are all blessed with exactly the same kind of creative capabilities as the universe: to bring into being that which has never existed before—from a child's drawing to a new technology—and everything in between.

We were born *creators*, not merely *created.*

Matthew Fox argues that "Creativity is seen as a spiritual, inwardly-driven activity, directly influenced by a Higher Power, or God. That is the ultimate in inspiration for me: to know I have "permission" to be creative and to be a creator too."[78]

Spirituality is seen as a pathway to God. When we see science in the new light of purpose of unity, of Creative Connecting, we also find a pathway to "God."

When we say "God," we are not necessarily referring to a being even remotely like the God described in the Bible, the Koran, the Iliad, the Bhagavadgita, or any other scriptural account of the supernatural. We are talking about the natural impulse to Creative Connecting that we have demonstrated exists throughout the universe. An impulse that, like a deity, we cannot explain. An impulse that is as important to us as any deity is to any religion, and that is surely worth "reverence" in the same way traditional religions command "reverence" for their particular deities.

This pathway has been followed since the universe began. It reveals a universe of creative intelligence that spawns offspring from subatomic particles to human beings, to carry out its purpose.

Science and spirituality do not merely co-exist, but they support and reveal and sustain one another.

Nature, it has been said, has no way of guiding "oughts" or "shoulds" or revealing ethics or morality. What do we say then of the way atoms or cells combine to create something

78. Matthew Fox, *Creativity: Where the Divine and the Human Meet* (New York: Tarcher/Putnam, 2002).

new, or the way the cells in our bodies collaborate or ecosystems of mutualism develop. *Nature at its deepest level is a model of morality.*

Testing this kind of model of collaboration, Niles Lehman of Portland State University looked at early self-replicating molecules and pitted selfish self-repairing molecules against collaborating molecules. The collaborating molecules won out and showed that they could create even larger networks of cooperating, mutualistic, complex molecules. They showed that real molecules can do this.[79] Nature strives to achieve creative collaboration.

We have designed institutions that teach conformity, competition and conventionality—not creativity! This limits thinking in most situations to the one "right answer"! This actually suppresses imagination and creativity. This does produce stability and predictability—for a time. But remember the cells that grew so successfully by replicating that they started to crowd and compete for resources so that, over time, conditions demanded a radical change. That change required a shift from competition and even combat for survival to a step into a world of collaboration and mutualism. Perhaps, now is that time with people. Today our global growth has brought about many problems; old solutions simply make things worse. We can learn from nature—from the universe.

And perhaps the old foundations of normative science and engineering and mathematics (that can't express creativity) might not be the proper focus for our current emphasis in education.

We might do much better to devote resources to reigniting our inborn creative talents, to innovating and even imagining new ways to get along with people who are different from us. We can start by cooperating with people who look

---

79. *Nature*, DOI:10.1038/nature11549

different from us. All people have the same human needs for mutual respect, life, safety, food, healthy children and vital things that enhance life rather than denigrate it or take it.

If we look at the intractable economic quandary facing the U.S. and other countries today it is fairly clear that attempting to return to the past, regenerating manufacturing for example, will not take us into the future. Simply "fixing" education will not transform our economy.

There is a solution. We can take advantage of our inborn capability for creativity—a capability now being ignored or even suppressed in our schools. It will take very little retooling to incorporate creative thinking into our educational system. Much less, in fact, than redoing math education.

This would be a revolution!

It would transform our economic system as innovation becomes widespread. In every economic study innovation proves to be by far the most powerful driver in any economy. Innovation comes from prepared minds. Mountains of data verify that just about everybody can learn to innovate! Yet we continue to teach conformity.

We must have the imagination and the will to change our educational system.

We must rethink how we think! This is what Einstein meant when he said, "We cannot solve our problems with the same thinking we used when we created them."

It is the only answer to creating a new and better future.

Our emphasis on the creative might pay other big dividends—like creative peace. We could give it a try, even at home between political parties and cultural divisions.

Perhaps we will recognize our replicative vulnerability and demand that we open our minds to our great potential for creativity and a mental and spiritual awakening.

This, after all, is what the god we describe here, nature and science combined, is now telling us.

And on the moral side, the universe provides us with a compass to guide us on our path: our moral and spiritual compass. It guides the carbon and oxygen atoms and the rest. It points the way to the good. It helps restructure after the bad has happened. The bad will always be mixed in with the good, but will never overwhelm it. The bad provides room for a new good.

Recall that connecting is the source of our pleasure and disconnecting the source of our pain.

Isolation removes us from life, just as love connects us. Is this the physics of love? What is the physics of deep spirituality—our sense of connection with the universe, with one another, and with love itself?

Up to now we have spoken in thinking terms, of cognition. What about emotion—feelings?

We ask you once again to imagine, to recapture an experience of a time when you felt at one with the universe. Perhaps at church. Perhaps standing on a beach at sunset holding the hand of a loved one. Perhaps it was captured looking out over the Grand Canyon, or just looking at a candle flame. Perhaps it was hearing a sound of music that transcended time, or gazing at a great painting. Imagine one of those transformative moments when you were wrapped in that kind of joy that defies words.

You were connected with everything, everywhere—beyond time or space.

We can talk about what we have observed, about what we believe. We can talk about what living a creative life can be about, and how working together creatively could make things better. In the end, what it is really all about is what you individually feel and cannot describe.

And so we have reached the end of our journey together into the depths of science and, we hope, into a bit of the profoundest spiritual longings of the human heart.

This way of thinking is new and different. We are asked to recall that we started this crossing with a deep distrust of long-accepted seeming "facts" of science and definitions like that from the Oxford "world's most trusted" dictionary that flatly says in its first definition of entropy:

> "1. PHYSICS, a thermodynamic quantity representing the unavailability of a system's thermal energy for conversion into mechanical work, often interpreted as the degree of disorder or randomness in the system: the Second Law of Thermodynamics says that entropy always increases with time.

We hope by now you have seen that this concept was in great error and in fact is the reverse of what actually happens in nature. Nature instead actually continually tends to bring everything together. What experimental evidence showed was that order and disorder *complemented* each other; that is, the "disorder" continually made room for new "order."

We took Einstein's words to heart: "If the idea is not absurd there is no hope for it." We reconsidered his special theory of relativity. By doing so, we combined both solutions of his relativity equation and revealed the complementary nature of causal determinism of classical physics ("past cause") with unknown "possibilities" of "future pull" and the unpredictable possibilities of quantum physics: the "brainstorming" of the universe. By uncovering the creative process, quantum mechanics can be understood!

Yet we are always on the brink of learning more. The noted philosopher of science, Karl Popper, sharply separates truth from *certainty.* He has written that what we think of as scientific knowledge "consists in the search for truth," but it "is not the search for certainty ... All human knowledge is fallible and therefore uncertain."[80] We have been in a search for what we believe is real, is truth, but the search will continue and new truth will emerge. We hope we have taken some new steps along the path.

Our unique experience and research in the field of creativity allowed us to recognize the unique complementary dynamic involved. As we have discussed, in the creative process, once a goal has been determined one explores many, many possibilities. The next step is to winnow those down or prune them to those which can fit and contribute to the environment—the ones that can work. Imagine, then judge.

This seems to be exactly what we have seen in nature: a continual creative process.

Everywhere we look nature universally practices a Creative Connecting process!

Now we ask you, what would be more simple and beautiful? What would make more "common sense" than a universal creative connecting process? A process so powerful that it has created beings with the abilities to do the same thing with their own minds?

Saint Thomas Aquinas said, "Since human beings are said to be in the image of God in virtue of their having a nature that includes an intellect, such a nature is most in the image of God in virtue of being most able to imitate God."[81]

"I say the Intelligent Design-evolution debate misses the point. It trivializes God and it trivializes science. The universe is like the hand of God. The world is God's body."[82]

We are not just creations but creators! Whether it is in making a lovely dinner for the family or a great work of art or

80. Karl Raimund Popper, *In search of a better world: lectures and essays from thirty years* (New York: Routledge, 1992) p.4.

81. Thomas Aquinas, *Summa Theologica* Ia q. 93 a. 4 http://www.sacred-texts.com/chr/aquinas/summa/sum103.htm

82. Michael Dowd, *Thank God for Evolution* (New York: Viking, 2007).

a new technology or a new friend, in the smallest and greatest ways we are pulled by a higher calling.

And leaving all that science stuff aside what have we been taught—across cultures, across generations, across religions?

We have been taught to empathize with, be compassionate with, understand, respect, value, care for, and love one another; to connect with, collaborate with, to cooperate with and work together: "To do unto others!"

Now we know this means to follow the same revelations that have flowed from the common sense elements of many religions. They have been revealing the same laws as now finally are also uncovered by a new look at the cosmos and its deepest nature and a new science.

There are consequences for breaking the law.

To act out of anger or fear or distrust is just simply breaking this most fundamental law of Creative Connecting. If plants did not act mutualistically, with benefit flowing on all sides, we would not have healthy ecosystems on land or in the oceans. They follow nature's laws of connecting and sharing differences—and flourish. To not do so is sad and painful.

We have the capability and calling to continue this work in this beautiful creative universe.

Just imagine what it will be like if we decide to work with this fundamental force of the universe, rather than against it; if we wake tomorrow, put our judgments aside, act individually and come creatively together to make this world work for everybody.

Not just to imagine . . . but to live out Creative Connecting,

To experience the power of LOVE!

APPENDIX ONE

# PERSONAL ACKNOWLEDGMENTS

Others that have contributed mightily include Barbara Marx Hubbard, Beth Ames Swartz, Bob Schwartz, Jean Houston, Larry Wilson, Arnold and Selma Patent, Greg Zlevor, Tom McNamee, Toni Morrison, Dr. Sanford Danziger, Donald Deptowicz, Bob Elmore, Shayla Roberts, Dr. Carol Marcus, Tony Baker, Jerry Marlar, Joyce Abbott, Mike Mazaika, Rev. Michelle Medrano, Robert Keim, Steve Zylstra, William Little, Bill Quigg, Gail Larsen, Dr. Frank Kinslow, Rev. David Wilkinson, Sid and Bea Parnes, Dr. Helene Wechsler, Dorie Shallcross, Rollo May, Bill J.J. Gordon, Abraham Maslow, Anthony Pozsonyi, Ing. Angel Sanchez, Marily Apodaca, Ing. Othón Canales, Gabe and Susan Heilig, Michelle Jarman, Brenda Ringwald, Tahdi Blackstone, Willis Harman, Paul Wray, Brian Swimme, Lynn McTaggart, Bruce Lipton, the Institute of Noetic Sciences Creative Education Foundation and our editors, Kim Kressaty and Luis Granados.

Profound influencers include:

Max Planck, David Bohm, Ernst Mach, Michael Dowd, Paul Davies, Margaret Mead, Dr. Karl Menninger, Werner Heisenberg, L. Von Bertalanffy, Theodosius Dobzhansky, Humberto Maturana, Deepak Chopra, George Gurdjieff, Herbert Spencer, Francisco Varela, Will and Ariel Durant, Oliver Reiser, Norbert Wiener, Henri-Louis Bergson, Jacques Monod, John A. Wheeler, Victor Frankel, Pierre Teilhard De Chardin, Stuart Kauffman, and far too many others to name here.

Not to be forgotten are our four adult children, who have supported us as we have been A.W.O.L. riding a rollercoaster during this time, Michelle and Alex Jarman and Bob and Patrick Land. We hope now to become the grandparents their children have heard about. And finally, Shirley Hudson, who kept the home fires burning.

We know we have forgotten more than a few. Forgive us please.

APPENDIX TWO

# CURRENT ENTROPY DEFINITIONS

## BOLTZMANN'S DEFINITION:

The second law, Boltzmann argued, was thus simply the result of the fact that in a world of mechanically colliding particles, disordered states are the most probable. Because there are so many more possible disordered states than ordered ones, a system will almost always be found either in the state of maximum disorder – the macrostate with the greatest number of accessible microstates such as a gas in a box at equilibrium – or moving towards it. A dynamically ordered state, one with molecules moving "at the same speed and in the same direction," Boltzmann concluded, is thus "the most improbable case conceivable... an infinitely improbable configuration of energy."

Ludwig Boltzmann, *The Second Law of Thermodynamics*, p. 20.

## ENCYCLOPEDIC AND DICTIONARY DEFINITIONS

First articulated in 1850 by German physicist Rudolf Clausius (1822–1888).

Entropy is a physical quantity that is primarily a measure of the thermodynamic disorder of a physical system. Entropy has the unique property in that its global value must always increase or stay the same.

This property is reflected in the second law of thermodynamics. The fact that entropy must always increase in natural processes introduces the concept of irreversibility, and defines a unique direction for the flow of time.

*The Gale Encyclopedia of Science*, January 1, 2008.

The concept, first proposed in 1850 by the German physicist Rudolf Clausius (1822–1888), is sometimes presented as the second law of thermodynamics, which states that entropy increases during irreversible processes such as spontaneous mixing of hot and cold gases, uncontrolled expansion of a gas into a vacuum, and combustion of fuel. In popular, nontechnical use, entropy is regarded as a measure of the chaos or randomness of a system.

Britannica Concise Encyclopedia, 1994-2008.

According to the second law of thermodynamics, during any process the change in entropy of a system and its surroundings is either zero or positive. In other words the entropy of the universe as a whole tends toward a maximum.

The Columbia Electronic Encyclopedia, 2007,
Columbia University Press.

en•tro•py

1. For a closed thermodynamic system, a quantitative measure of the amount of thermal energy not available to do work.
2. A measure of the disorder or randomness in a closed system.
3. A measure of the loss of information in a transmitted message.
4. The tendency for all matter and energy in the universe to evolve toward a state of inert uniformity.
5. Inevitable and steady deterioration of a system or society.

[German *Entropie* : Greek en-, in; see en-2 + Greek *trop*, transformation; see trep- in Indo-European roots.]

The American Heritage Dictionary of the English Language, Houghton Mifflin Company, 2009.

[E]ntropy is a measure of the disorder of a system. On average all of nature proceeds to a greater state of disorder. Examples of irreversible progression to disorder are pervasive in the world and in everyday experience. Bread crumbs will never gather back into the loaf. Helium atoms that escape from a balloon never return. A drop of ink placed in a glass of water will uniformly color the entire glass and never assemble into its original shape.

Entropy as a measure of disorder can be shown to depend on the probability that the particles of a system are in a given state of order. The tendency for entropy to increase occurs because the number of possible states of disorder that a system can assume is greater than the number of more ordered states, making a state of disorder more probable. For example, the entropy of the ordered state of the water molecules in ice crystal is less than it is when the crystal is melted to liquid water. The entropy difference involved corresponds to the transfer of heat to the crystal in order to melt it.

Lawrence W. Fagg, "Entropy," in Wentzel Van Huyssteen et al., *Encyclopedia of Science and Religion* (MacMillan, 2003).

APPENDIX THREE

# CREATIVITY AND PROBABILITY

---

Let's take a few minutes and think about this question of creativity, as well as this very interesting question of this thing we call "probability."

Imagine that you work for an organization, in a very cushy job. You're in your big recliner chair and the boss comes in with a request. "We need some ideas. Our division in Botania has found huge reserves of iron. They have lots of gold too—so no limit on what you can spend. So what new products could they make with all that iron? You've got all the time and money you want. Heck, you could come up with diamond-studded elephant belts. Go to it."

So pull out that five-year-old and start to dream. Anything is possible. At first, you're stymied. You look around your office. Iron desk? Iron chair? Too mundane? But you know you need to consider *everything* just to clear your mind. So you start in. Before you know it, you've started to spin stainless steel fabrics and make do-it-yourself homes. Nobody told you that you couldn't mix the iron with other things. Why not make iron woven into glass?

And so hundreds of ideas flow. As Einstein said, "Imagination is more important than knowledge. For knowledge is limited, whereas imagination embraces the entire world."

At some point you think you've got enough. Now it's time to sort. Which might be the good ideas—the ones that are practical, the ones you could actually make and people would want to buy? You set up some piles: terrible, no good, maybe, pretty good, really good. And you start to sort away.

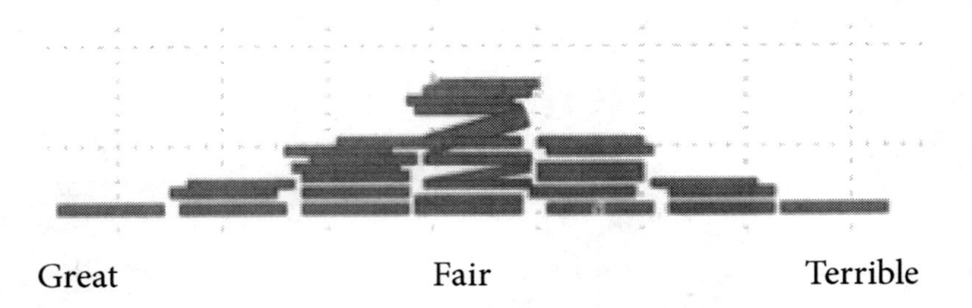

Great Fair Terrible

Gee, you've seen this kind of pattern before. It's the famous bell-shaped curve. It was in your class in probability. They taught you that everything in large numbers comes out in these patterns—in gambling, in life insurance, in marketing.

When you stop to think about it, it even describes what you're about to do right now as you sit here. You look out at an unlimited future of possibilities. You could continue to sit here. You could go over to your desk for a drink, or go down to the rest room or go home or go get on a plane to Paris or ... just about anything. But there are probabilities. If you made piles of lists of what you might do they'd end up just like the piles you just finished making. How about that?

Sitting there with this book, you are in a sea of probabilities.

Well, now let's make a big left turn into the world of meaning; that is, why do things work the way they do? For a very long time the ancients believed that the universe was organic. Things had their own *nature*. Rocks fell because it was their nature. Likewise trees grew because it was their nature. Birds flew and so on. Copernicus and Newton changed all that *radically*. They showed that the universe was a big, gigantic machine. They were relentlessly deterministic. If you knew the past you could accurately predict everything that would happen in the future. Mathematics was God's law. From the 1600s to the 1920s this was the belief system.

In the 1920s, quantum physics destroyed determinism forever. This theory has been verified more than any idea in the history of science. As we have seen, one of the most fundamental findings was that every solid particle in the universe exists as both a real piece of matter and as a "wave." Let us be as clear as possible about this finding that scientists to this day call "impossible." If you look at an electron, for example, a part of an atom, one way it appears as a solid particle with mass and it occupies a particular place in space. If you look at it another way it appears everywhere, with higher or lower probabilities of finding it. It is not a "thing" it is ethereal, just a mathematical formula.

Scientists accept this as the principle of "complementarities." These two aspects of nature somehow complement each other. The mathematical formula is called a "wave" because it takes on the shape of a wave when the math is graphed. It looks like the bell-shaped curve repeated, which looks like waves.

We humbly want to suggest that the particle carries with it the past history of the object, an atom or a person, and the wave carries its possible – its probable – futures. The object "tames" or guides the wave with its past history, turning possibilities into probabilities. It introduces enough order into the disorder of the possibilities so that new order can emerge.

The wave of possibilities and the particle of reality combine in the creativity of the present. They continually bring about the new, the different sets of connections that fulfill the connecting, unifying purpose of the universe.

That pile of ideas resembles something we have often seen called a "normal distribution" of probabilities. The "bell-shaped curve":

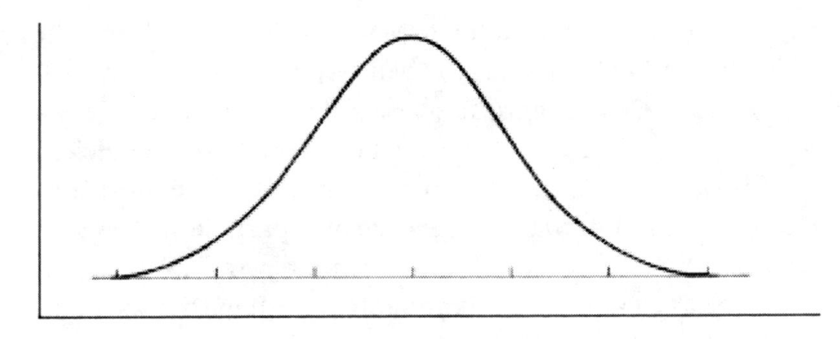

Events happen in probabilistic ways; they are not completely random and they are also not predictable or deterministic. How does the seeming order of probability come out of the chaos of randomness? This has lingered as an unsolved problem in mathematics, science and philosophy for a long time.

In the 1650s the study of gambling uncovered the idea of probability. Over time evidence accumulated that such a phenomenon existed and could be depended on in predicting the course of events, at least enough for profit in gambling, insurance and the like. But there has never been a logical, formal theory of natural processes to explain probability. We propose that the creative connecting processes are at work.

The creative process combines the infinite imaginative possibilities of disorder with the order of decision choices.

In special relativity, the overlay of the order of past determinism (the positive solution) combining with the randomness of future pull (the negative solution) results in guidance of narrowed order moving forward.

Likewise, in quantum physics infinities are renormalized.

Plant hybridization produces normal probabilistic distributions.

We have gotten to thinking about probabilities and the wave as "Casper the invisible accounting ghost in the dice." When you shake the dice, you rattle Casper around but he is keeping track of the waves and riding along as they fly through the air. When they land he very quickly figures things out so that for sure in the long run the odds will even out the way they are supposed to in that bell curve.

He is very busy. And accurate. And he has a big team—throughout the universe.

APPENDIX FOUR

# ABOUT THE AUTHORS

**George Land**, General Systems Scientist, discovered Transformation Theory. He is the author of *Grow or Die, the Unifying Principle of Transformation*, published by Random House, Dell, John Wiley and Sons and the Creative Education Foundation. It was a Main Selection of the Saturday Review Book Club, and was submitted by Random House for the Science section of the National Book Award and the Pulitzer Prize. He co-authored *Breakpoint and Beyond* with Beth Jarman, published by Harper and Harper Perennial. George Land was elected a Senior Fellow of the University of Minnesota, a Fellow of the New York Academy of Sciences and was inducted into the Creative Education Foundation's Creativity Hall of Fame. George's hobbies include reading, writing, anthropology, archeology and hiking—from our truck.

**Beth S. Jarman, Ph.D**. Beth's broad-based experience includes six years as an educator, cabinet secretary to two governors in two states, a state legislator, president of Leadership 2000 and CEO, The FarSight Group. Her publications include *You Can Change Your Life by Changing your Mind* and *Breakpoint and Beyond*, She is a Founding Member of Women Executives in State Government and the Founder of the Phoenix City Club.". She is a recipient of the following awards: YWCA & Utah Housing Finance Agency and Phoenix City Club Distinguished Service Awards; Abrelia Clarissa Hinckley Graduate Fellowship; and the Distinguished Alumni Award from the University of Utah, 2012. Beth's passion is gardening, where she can see and participate in nature's amazing creativity.

More complete biographies for George Land and Beth Jarman can be found in *Who's Who in America* and *Who's Who in the World*.

**Contact the Authors:**

George Land: geozeno@cs.com
Beth Jarman: bethjarman@cox.net

APPENDIX FIVE

# CLIENT ACKNOWLEDGMENTS

---

Our gratitude to the many organizations with whom we have collaborated and from whom we have learned so much (such as Leadership 2000, and The FarSight Group).

## PRIVATE SECTOR

3M
3i (UK)
Allied Domec (Mexico)
American Hospital Association
American Can Company
American Medical Systems
Atlantic Richfield Sinclair
AT&T
Grupo Azteca (Mexico)
Banner Hospitals
Bell Labs
Block Drug Company
Branch Bank and Trust Company
Bristol-Myers Company
Carter-Wallace Inc.
CIBA Geigy
Coca-Cola
Credit Union National Association
Creative Education Foundation
CUNA Mutual Insurance Company
Daimler Chrysler
Delphi Dow Chemical
Dow Corning
E.I. DuPont de Nemours & Co.
Edison Electric Institute
General Foods
General Mills
General Motors
Gilbane Building Corporation
HealthNet
Herman Miller, Inc.
Hewlett Packard
Hyatt Hotels
InterMedics Orthopedics
Johnson & Johnson Medical
Johnson Wax Professional
Kimberly Clark
Liberty Life Insurance Company

Midamerican Energy Corporation
Mitsui Corporation (Japan)
Monsanto Corporation
Moore Corporation
MSS Technologies
Nestle
Pepsi Cola
New England Nuclear
Novell
Owens Corning
PEMEX (National oil company of Mexico)
Peñoles (Mexico)
PDAVESA (National oíl company of Venezuela)
PriceWaterhouseCoopers
Procter & Gamble
R. J. Reynolds Foods
S. C. Johnson & Sons
Science Applications International Corporation
Schuster Center for Professional Development
Spinetek
SulzerMedica Corporation
Takanaka Construction (Japan)
Time Inc.
Topan Moore Corporation (Japan)

## PUBLIC SECTOR

City of Tempe, Arizona
Defense Evaluation &
Research Agency (UK)
International Monetary Fund
Los Angeles Police Department
Organization of American States
Space and Naval Warfare
Systems Center
United States Office of Personnel Management
United States Department of Commerce
United States Department of Energy
United States Department of Agriculture
United States General Services Administration
United States Patent and Trademark Office
United States Senior Executive Service
United States Federal Reserve
Ministry of Defense (UK)
SAT (Mexican Internal Revenue Agency)
Federal Electric Commission (Mexico)

## UNIVERSITIES

Stanford
MIT
University of Minnesota
St. Johns University
Macalester College
State University of New York

# YOUR TURN

Did you like it? Did you hate it? How would you critique the authors' argument? What aspects of Creative Connecting and the overlap the authors find between science and spirituality did you find most appealing?

Please share your thoughts with other readers at:

**www.humanistpress.com/natures-hidden-force-feedback.html**

# Index

## A

Addis, Donna Rose 94
algae 122, 123
aptamers 70
Aquinas, Thomas 148
Ash, Mary Kay 101
astrobiology 141
atoms 23, 37, 41, 44, 45, 50, 54, 75, 77, 78, 79, 85, 90, 94, 99, 113, 114, 143, 145
Aurelius, Marcus 77
Ayers, Joseph 70

## B

Bacon, Roger 24
bacteria 114, 123
bell-shaped curve 156
Bentley, Peter 70
Big Bang 11, 77, 113
Binder, Steve 104
biology 73, 117
bisociation 123
Blake, William 131
Bohr, Niels 69, 82, 93, 139
Boltzmann box 37, 42, 44
Boltzmann, Ludwig 7, 27, 30, 35, 38, 51, 63, 152
Born, Max 67
Brandon, Robert 72
Brown, Ali 104
Brownian motion 41
Bruno, Giordano 56
Buckner, Randy 95
Byers, William 25

## C

Cambrian explosion 78, 122
Campbell, Joseph 105
carbon 113

causality 86, 138, 139
cells 48, 60, 69, 70, 75, 79, 114, 115, 116, 117, 119, 121, 122, 143, 144
certainty 147
change 115, 116, 118
chaos 48
children 128, 132
Clark, Maxine 104
clash of cultures 119
Clausius, Rudolf 27, 29, 51, 152, 153
Clayton, Philip 14
collaboration 122, 123, 143, 144
complementarity 63, 69, 75, 93, 131, 139, 157
consciousness 94, 97, 99, 106, 141
content, context mistake 47
Copenhagen Interpretation 93
Copernicus 156
covalent bonds 122
Creative Connecting 33, 51, 75, 76, 83, 108, 110, 112, 118, 120, 121, 131, 141, 142, 143, 147, 148, 149
creative destruction 49, 50
Creative Education Foundation 77, 130, 132, 136
creative restructuring 49
creative thinking 135
creativity 10, 11, 12, 19, 53, 118, 119, 123, 125, 128, 129, 133, 134, 136, 137, 140, 144, 147, 155
  characteristics of 54

## D

Dalai Lama 3
Davies, Paul 14, 140
Dawkins, Richard 5
day-dreaming 130
de Chardin, Pierre Teilhard 34, 120
deity 3, 4, 40, 143
DeJoria, John Paul 103
Dell, Michael 101
Del Vecchio, Leonardo 103
determinism 61, 63, 66, 147, 156, 158
Diamond Jr., Arthur M. 49
Dirac, Paul 55, 68, 108
disorder 6, 7, 19, 24, 27, 28, 29, 30, 33, 34, 37, 38, 39, 40, 43, 45, 58, 59, 71, 74, 78, 89, 121, 141, 142, 147, 152, 154, 158
Divergent Thinking 130

DNA sequencing 118
double slit experiment 87, 106
dreaming 134

## E

Eddington, Sir Arthur 4, 8
Edison, Thomas 132
education 102, 128, 129, 133, 144, 145
Einstein, Albert 4, 8, 12, 14, 15, 24, 31, 41, 50, 51, 58, 59, 60, 64, 65, 68, 76, 87, 112, 122, 124, 126, 127, 133, 135, 137, 139, 145, 147, 155
electrons 26, 44, 60, 77, 81, 84, 85, 87, 89, 91, 92, 94, 96, 99, 157
Ellis, George Francis Rayner 73, 91
E=mc2 61, 63, 66
emergence 11, 26, 74, 75, 98, 118
emergent phenomena 54
emotion 146
entropy 4, 5, 7, 17, 23, 27, 50, 112, 137, 146, 152, 154
Epicurus 50, 59, 110
eukaryotic cells 117
evolution 117, 118, 122, 126, 140
existentialism 40
extinction 53

## F

Fantappiè, Luigi 34, 52
feelings 146
Feynman, Richard 18, 26, 56, 64, 82, 95
Foundation for Mind Research 134
Fox, Matthew 71
Fuller, Buckminster 133
fungi 73, 121, 123
future pull 66, 87, 91, 92, 96, 98, 99, 100, 101, 102, 147, 158

## G

Galilei, Galileo 24
gases 20, 28, 30, 51
genes 114, 115, 116, 117, 121, 123
God 4, 5, 7, 14, 18, 63, 111, 142, 143, 145
Golden Rule 76
gravity 41, 45, 52, 79, 109, 110, 111

Gray, Michael 73
Guilford, Dr. J. P. 128, 129

## H

Hawking, Stephen 34, 73
heat 28, 29, 33, 39, 50, 51
heat death 5, 32, 40, 45
Heisenberg, Werner 78, 85
Hesburgh, Theodore 102
Houston, Jean 134
Hubbard, Barbara Marx 126
Human Potential Movement 134
Huxley, Julian 126

## I

imagination 131, 132, 135, 145, 155
Imry, Yoseph 107
Industrial Revolution 38
infinity 55
innovation 125
Ireland 16

## J

James, William 111
Jarman, Beth 10, 21, 160
Jeans, Sir James 68, 141
Jobs, Steve 101

## K

Kamprad, Ingvar 104
Kauffman, Stuart 141
Kim, Kyung Hee 133
Koestler, Arthur 91, 123

## L

Laliberté, Guy 103
Land, George 9, 10, 21, 160
Law of Attraction 100
law of large numbers 90
Learning Innovation and Technology Consortium 124
Lehman, Niles 143

Lewis, William 46
lichen 123
Lovász, László 26
love 76, 111, 149

## M

mathematics 13, 23, 24, 25, 51, 54, 56, 68, 140, 144, 156
McShea, Daniel 72
McTaggart, Lynne 106
Meier, Debbie 102
Monod, Jacques 98
morality 143
Morin, Edgar 6
Morowitz, Harold 14
Murphy's Law 5, 28
mutualism 116, 117, 119, 122, 123, 124, 143, 144

## N

nature 8, 13, 21, 22, 25, 28, 30, 31, 33, 48, 51, 52, 53, 54, 56, 59, 61, 65, 67, 74, 76, 77, 78, 81, 82, 84, 87, 88, 94, 110, 112, 113, 114, 116, 117, 119, 121, 124, 131, 134, 135, 140, 143, 144, 145, 147
negative entropy 52
negentropy 52
Newton, Isaac 3, 28, 135, 156
non-determinism 62
noosphere 120
nucleons 90

## O

observation 93
observer 107
Ortega y Gasset, José 98

## P

Pagel, Mark 118
Parnes, Dr. Sidney 132
particles 81, 82, 84, 140, 157
Pauling, Dr. Linus 132
Pauli, Wolfgang 67, 71
plants 123
Plato 24
Popper, Karl 147

positive thinking 11, 12, 13, 83, 97, 99
postmodernism 40
pragmatism 111
Prigogine, Ilya 67, 69, 78, 141
probabilities 62, 78, 84, 90, 93, 94, 95, 96, 155, 156, 158
Pythagoras 135

## Q

quadratic equations 65
quantum indeterminacy 138
quantum mechanics 26, 29, 61, 62, 63, 64, 83, 84, 86, 93, 131, 138, 139, 147
quantum physics 12, 39, 55, 62, 63, 66, 72, 82, 83, 138, 139, 147, 156, 158

## R

randomness 63, 70, 85, 86, 89, 139, 153, 158
relativity 29, 60, 61, 63, 69, 75, 138, 147, 158
religion 40, 66, 76, 148
renormalization 56
replicative vulnerability 115, 123
Rifkin, Jeremy 4, 28
Roddick, Anita 103
Rouse, James 101
Rowling, J. K. 104
Rubi, J.Miguel 46

## S

Sagan, Carl 132
Sakellariou, Christos 70
Schacter, Daniel 94
Schrödinger, Erwin 41, 52, 94, 97, 108
Schumpeter, Joseph 49
science 3, 4, 5, 6, 7, 8, 10, 11, 14, 18, 19, 21, 24, 26, 28, 29, 30, 31, 32, 39, 40, 51, 54, 60, 64, 66, 67, 74, 76, 81, 84, 137, 138, 142, 143, 144, 145
second law of thermodynamics 4, 5, 19, 32, 34, 39, 40, 46, 50, 137, 146, 152, 153
sex 115, 116, 122, 123, 124
Shaw, George Bernard 101
Shenk, David 129
Speijer, David 73
spirituality 5, 8, 32, 40, 60, 67, 74, 75, 76, 78, 137, 138, 142, 143, 145

square roots 65
Stapp, Henry 52
Stern, Adi 107
subatomic particles 13, 54, 75, 87, 88, 96, 143
supernovae 54
symbiosis 53, 123
syntropy 47, 52, 112, 113, 141
Szent-Györgyi, Albert 31, 47

## T

Thomson, J. J. 37
Thornton, Joe 73
thought experiment 35, 51
transhumanism 126
truth 147

## U

Übercausalität 67
universe ix, 3, 5, 6, 7, 13, 14, 18, 21, 23, 24, 27, 28, 30, 31, 32, 34, 35, 38, 39, 40, 41, 45, 46, 51, 56, 59, 62, 63, 65, 68, 72, 76, 79, 85, 95, 96, 97, 99, 100, 109, 112, 137, 138, 141, 142, 143, 144, 145, 156

## V

Von Neumann, John 90

## W

Walton, Sam 104
wave/particle duality 82, 139
waves 84, 85, 86, 88, 89, 93, 95, 96, 97, 98, 99, 100, 107, 108, 139, 157
Weinberg, Steven 6
Wheeler, John 69, 94
Whitehead, Alfred North 25
Wilber, Ken 75
Winfrey, Oprah 104
Wozniak, Steve 101

## Z

Zimmer, Carl 72

CPSIA information can be obtained
at www.ICGtesting.com
Printed in the USA
FFOW03n2117240714
6467FF